BECOMING ARTIFICIAL

A PHILOSOPHICAL EXPLORATION INTO ARTIFICIAL INTELLIGENCE AND WHAT IT MEANS TO BE HUMAN

Danial Sonik and
Alessandro Colarossi

imprint-academic.com

Published in the UK by
Imprint Academic Ltd., PO Box 200, Exeter EX5 5YX, UK

Distributed in the USA by
Ingram Book Company,
One Ingram Blvd., La Vergne, TN 37086, USA

ISBN 9781788360203 paperback

A CIP catalogue record for this book is available from the
British Library and US Library of Congress

Get lost in this book,
in a way only a human can.

Contents

Preface

Artificial intelligence (AI) is a hot topic as we write this preface in 2020. To be fair, AI has been a hot topic for the past few decades, but it has only been recently (in the past five or so years) that many approaches in deep learning have begun to bear fruit, a bumper crop that we are still only struggling to harvest today. This is why there is so much consternation on all fronts as to how AI will affect what we today consider the "status quo" in human society and behavior.

This social concern (and dare we say it, dismay) is reflected in a number of alarmist journal articles that seem to have headlines preparing everyone for the worst. To take some of the most straightforward and panic-inducing examples: "Higher Paying White Collar Jobs Will Be Heavily Impacted By AI," "AI Will Transform 500 Million White Collar Jobs in 5 Years; Silicon Valley Must Help", "Is AI Going to Be a Jobs Killer?", and "Types of Jobs AI Will Eliminate by 2030."

Of course, we all know that the news thrives on panic and speculation. If it bleeds, it leads. News sites get paid by the click, and clicks only come from people who feel some sort of anxiety about the future and want to prepare for the worst. Similar types of doomsday pronouncements were made in the lead up to Y2K, and they all pretty much turned out to be a nothing burger (with the exception of a few inconvenient and embarrassing counterexamples, such as minor flight delays, some websites incorrectly referring to the new year as 1900, and, of course, a small contingent of babies born already 100 years old due to lack of updates to computer software running in hospitals).

Sampling the many books that glut this popular field, it appears as though they can be broken down into one of three camps. First, there are the "gee whiz" triumphalists (Max Tegmark calls them "digital utopians") who talk seriously and with a straight face about being able to upload consciousness into "the cloud" and gain a sort of digital immortality. Offenders in this genre include the absolutely brilliant (albeit, in our opinion, misguided) Ray Kurzweil (*The Age of Spiritual Machines*) and fellow Kool-Aid drinker Michio Kaku (*The Future of Humanity: Our Destiny in the Universe*). Then, there are the techno-apocalyptics (Tegmark refers to them as "luddites") who think that AI is setting humanity up for a disaster it will not recover from. Folks in this category include James Barrat (*Our Final Invention*) and the late Stephen Hawking, who was quoted as saying at a web summit technology conference: "Unless we learn how to prepare for, and avoid, the potential risks, AI could be the worst event in the history of our civilization. It brings dangers, like powerful autonomous weapons, or new ways for the few to oppress the many. It could bring great disruption to our economy." The third camp in the AI biblioverse is authors who take a measured approach—AI is not going to solve all of our problems, but nor is it going to bring us to the brink of destruction. Max Tegmark (*Life 3.0*) and Nick Bostrom (*Superintelligence: Paths, Failures, Strategies*) fall into this camp.

This book chooses a fourth way, comprising both pragmatic and philosophical components. On the pragmatic side, AI will change the world—no question. On the philosophical side, the changes that will be ushered in by AI will have NOTHING to do with AI's ability to have or gain sentience. There are simply too many features of human consciousness and sentience that cannot be described in algorithmic form, and therefore are not amenable to any computational approach, even though computational approaches seem to be the only game in town when it comes to building artificial intelligence.

In order to pursue this fourth way, we aim to tackle a significant subset of the most pressing problems for humanity to consider (and possibly solve, if they are tractable) with respect to this topic in the form of both fictional and non-fictional treatments. Fields such as elder care, warfare, journalism, academic knowledge production, and the legal profession are all poised for a radical transformation due to relatively recent advances in AI. However, the AI methods that have contributed to this massive transformation hold little promise or insight into the holy grail that seems forever to recede into the distance — how to generate an actual instance of consciousness using nothing more than silicon chips and electricity.

In the first chapter, we provide an abridged history of AI, highlighting the most important parts relevant for our key argument. After providing a brief history of this discipline in computer science, we hope to take you on a journey from the beginnings of AI to its current state today and the eventual form that it might have in the next 50 years.

Then, in Chapter 2, we recount and elaborate philosophical arguments originally formulated by Merleau-Ponty on the irreducible importance of embodiment in the development of intelligence and the sense of self. In doing so, we will ponder the question of just how feasible "abstract consciousness" is given the seeming primacy of our bodies in how we model and interact with our world.

For Chapter 3, we zoom in on the mind itself and canvas a number of approaches (all currently unsuccessful without major, and as yet unknown, modifications) that attempt to model the mind. We conclude that the mind itself appears to be an intractable problem and unlikely to be instantiated in an AI given the state of the art in machine learning and machine intelligence as we write this in 2020.

Notwithstanding the skeptical conclusions in the previous chapter, "The Changing Nature of War" describes the ways AI will be applied to warfare. This application opens up the very real possibility of machines taking human lives according to their programming and prompts an ethical discussion of how to deal with this future.

In Chapter 5, we look at the idealistic portrayal of AI and robots in film and highlight the "uncanny valley effect" that always seems to arise whenever AI is represented in fiction. We will argue that the uncanny valley is not merely an artifact of how well the AI is portrayed, but rather is an unbridgeable gulf in principle between what we think robots can do and what they can actually do. We also highlight how the "uncanny valley" and its "imagination of the artificial" also evolve in human history.

Chapter 6 talks about the rise of "emotional support robots," arguing that while such AI creations may not have any inner life as such, they nevertheless get imbued with human characteristics by their human "lovers." This, in turn, will provide a good explanation for why humans constantly seek out a form of "sentience" in their AI/robotic creations, even though there are no grounds for actually thinking that their little mechinic companions in fact do possess sentience—rather, we give them the sentience by including them in our social rituals.

"Alone at the Water Cooler," Chapter 7, looks at how the world of work is going to be transformed by AI, in particular in the areas of speech and language processing. Current text generation applications such as GPT2 seem to be able to generate block after block of superficially intelligible text given huge libraries of input (i.e. scraping the internet). However, will their ability be enough to displace people who rely on the written word for their livelihood? And who would be the consumers of such computer generated text in the first place?

In "Do As You Are Told", we argue that, ultimately, any AI must of necessity and by design be composed of a set of finite and discrete instructions (whether these instructions are human generated or else created by the computer on a path of "machine learning" does not matter). This is ultimately because no machine can be coded with either will or desire. Moreover, we have no adequate theory or algorithm by which we would be able to encode these mysterious human traits.

"The Devil Made Me Do It" looks at the question of legality and ethics in a world shared by humans and AIs. While we will maintain that AIs will never have the sentience possessed by humans, it might nevertheless be possible for them to be party to legal actions in the same way that corporations currently are. This could pose massive problems for a justice system that changes at a comparatively glacial pace.

The final chapter, "Incomplete Transfer (Upload or Die)", looks at questions of identity and consciousness that are raised when entertaining the idea of uploading consciousness. It then questions its desirability given our reliance on embodiedness as highlighted in Chapter 2.

A narrative thread runs throughout the book laying out some of the very real changes that are on the horizon, integrated with essays that describe both the hopes and limitations that current AI approaches have. Taken together, each of these chapters will provide an additional set of considerations that work in favor of the following claims:

Thesis 1

"Sentient" or "Qualia Aware" or "First-Person Introspective" or… substitute whatever you think needs to go within the "scare quotes" in the preceding text. For the purposes of reference in this text, let's call it X, that ineffable something Descartes located in his very simple argument: I think, therefore I am. (For if I was not, who would be able to doubt that I actually was?)

Let's also say this: If you are reading this text right now and you know what it is like to read this text right now, then you possess "X". It's THAT that makes you reading these words, something that is not encapsulable within the current limits of AI, in practice OR IN PRINCIPLE. This is a very strong thesis we will attempt to defend on all fronts in the upcoming chapters.

Thesis 2

Notwithstanding the truth of Thesis 1, AI on its current trajectory is poised to shatter and disrupt life for ALL humans within the next 20 to 50 years. This is not because it will actually become sentient *à la* Skynet in *Terminator*, but rather because we will have built it into so many places within our social fabric that it will be impossible to escape. This brings with it the danger of, for lack of a more expressive way of putting it, "shitty AI" — AI that has been rushed out the door due to some perceived corporate advantage combined with the more common apathy most technology companies now practice with respect to quality control. (Rather than hire testers to test the software before release, why not let our customers do it instead? What else can they do? Use another operating system/chipset/platform?)

The pace at which many human agencies and innovators have embraced the evolution of the current set of AI techniques makes the possibility of a "self-aware" AI much less real, but at the same time it makes the possibility of an AI army controlled by a single political interest much more viable. Just imagine: the same handful of tech companies that already control the digital economy can now do so with far less input and more automation—what will the future of politics, society, and day to day opportunities look like in a new world dominated by AI controlled by entrenched political actors? And what is to stop companies or governments from releasing AI that subtly tilts all interactions in favor of their own interests?

Thesis 3

Notwithstanding the truth of Thesis 2, AI is only as good as the information that you provide it with. If you give AI bogus information, it will reach bogus conclusions. This is where we arrive in the final chapters, with some strong possibilities (as well as strong ways of acting to avoid such possibilities) for AI to control the criminal justice system, your potential future life after death, and the ultimate "retainment" of all your memory contents after death in some cloud mechanism that copied your every brain synapse firing without ever getting to the heart of YOU. YOU know who you are.

Chapter 1

A Brief History of Artificial Intelligence

The current state of the art in AI has found a number of practical uses in areas like analyzing consumer purchase histories, guiding marketing strategies, and providing voice-activated virtual assistants. AI has even found its way into corners of human endeavor where we would traditionally be wary of allowing anything but human experts. In the medical domain there have been machines that provide expert medical diagnoses since at least the mid-80s. CADUCEUS, for example, was a system that could provide diagnostic inferences from a massive knowledge base of known diseases and their symptoms (Winston & Prendergast, 1986). SHYSTER was created in the early 90s as a proof of concept to determine outcomes in Australian case law (Popple, 1993). Yet, it remains a work in progress, and a work that seems to always promise more than it can deliver, even as every latest iteration brings us closer to the ever-vanishing horizon of perfect. From this inductive view, we can see at least for the time being that AI's developmental history is still being written. If there is an end in sight, it is almost certainly a mirage. The finish line is an elusive, moving target in full view of a fascinating journey encompassing a modern-day timeline stretched beyond six decades. Are we there yet? Not all the way. And perhaps never, because the changes in AI research constantly make us re-evaluate the questions of what it is to behave intelligently, not to mention the questions of what make us characteristically human.

The quest for an AI, or more accurately an AI that is "general" in the same ways that our human intelligence is general (an "artificial general intelligence," or AGI, as most practitioners of the field have decided), is a science-driven expedition in search of technology's holy grail—a machine that can perform just as well if not better than a human being in all of the areas of endeavor that require cognition of some form. In some areas, machines already have us beat, hands down. If you are a person who follows these developments (and I think you are, because you have this book in your hand), you certainly already know that Google's DeepMind algorithm has out-stripped all potential human capacity for success in zero-sum games such as chess and Go. Chess supremacy by AIs was arguably achieved with the open-source engine Stockfish against grandmaster player Hikaru Nakamura in 2014. After besting this top world ranked human, Stockfish went on later to be humbled in ignominious defeat by DeepMind's AlphaZero a mere three years later, ushering in what has become a time of supremacy for chess playing "entities"— humans are simply no longer competitive in this arena.

Go, forever the bastion of zero-sum game complexity and ridicule by AI skeptics and naysayers, was the next to fall. The argument against AI systems playing as well as humans in this ancient game came down to a matter of combinatorics. While a brand new chess game will have 20 valid moves for each side from the starting position, Go has (depending on the size of the board) anywhere from 169 to 361 opening moves. This vast difference in the opening character of each game translates into a correspondingly mind-boggling increase in the number of positions possible on the board 10, 15, or 20 moves in.

There still, however, seem to be some gaps. Can an AGI write a novel or conduct an intelligent conversation? Or perhaps create novel works of art, something like a cyber-Picasso or an iWarhol? The very idea of a machine that would be able to communicate or interact with a human interlocutor in a creative and unpredictable way is an old one. The most common benchmark for the origins of what we now know as modern-day AI science dates back to the mid-1950s, but the accumulation of knowledge and enterprise directed at under-standing how humans think is many centuries older—indeed, thousands of years in the making. No less than Aristotle, sometime during the 300s BC, was preoccupied with finding a route to advanced thinking, despite his never having watched *Star Trek* episodes or *Blade Runner*'s silver-screened AI-engineered replicants to put such ideas into his head. Aristotle's pioneering of the syllogism, arguably humanity's first attempt at creating an "automatic pilot" for reasoning and deducing correct claims as output given symbolic input in the form of major and minor premise, anticipated today's theorem proving machines. Aristotle unwittingly became an AI advo-cate, describing in his collection of six writings on logic—named *The Organon* by his followers (the Peripatetics)—something known as syllogism, a method of formal, mechanical thought combined with his theory of knowledge (McKeon, 1941; Giles, 2016). Likewise, Greek mythology had its place in the early manifestations of elementary AI. The myths of Hephaestus and Pygmalion mention intelligent robots and fabricated beings, such as those known as Galatea and Pandora (McCorduck, 1979).[1]

As centuries passed, incrementally modest developments were espoused by the great minds of their respective times. They offered theories, calculations, and other writings that would comprise a foundation of knowledge suitable for mechanizations and advanced technologies. These other early AI pioneers included Porphyry of Tyros, who in 260 AD produced a work known as *Isagoge*, which categorized knowledge and logic (Russell & Norvig, 2003); Geber, who around the ninth century conceived what was known as Takwin, an Arabic alchemical theory that postulated the artificial creation of life in the laboratory, up to and including human life (O'Connor, 1994); and Sir Francis Bacon, who in 1620 came out with his empirical theory of knowledge. For good measure, Bacon proposed a concept known as inductive logic in his work titled *The New Organon*, an obvious homage to Aristotle's centuries-old forerunner *The Organon* (Bacon, 1620/2000).

There were outlandish developments along the way as well, at least in the hindsight of scientific minds who embrace pragmatism in place of mythology, astrology, alchemy, and reading bark. Under the influence of the scholastic tradition

---

[1]   Make note, though, that the word "robot" didn't become part of the scientific vernacular until the early twentieth century. That was courtesy of playwright Karel Capek's play *R.U.R. (Rossum's Universal Robots)*, which opened in London in 1923 and is believed to make the first English language mention of the word 'robot' (McCorduck, 1979). But it was clear, even those thousands of years ago, that man already was conceiving some sort of artificial or mechanical humanoids infused with human knowledge.

that Aristotle had set in motion, Catalan theologian Ramon Llull proposed in the fourteenth century an "ultimate general art," a mechanical method by which one could combine topics and ideas from religion and philosophy to arrive at novel insights in both. In 1275 he formulated a tool he termed the Ars Magna, which combined concepts mechanically predicated on the Zairja, an Arabic astrological tool. To Llull's credit, his formulations more than three hundred years later would be further developed by Gottfried Leibniz (McCorduck, 1979). Hardly known in the modern world (intellectual giants such as Leibniz and Turing seem to have eclipsed Llull's originality), Llull's conception of "automated creativity" is almost certainly a key idea underpinning of all AI and AGI research today.

Circa 1500 saw Paracelsus, a Swiss physician, alchemist, and astrologer, make claims that he had manufactured an artificial man through a combination of magnetism, sperm, and alchemy. To be fair, we concede that Paracelsus also exercised a more rational side, considering he was acclaimed as the "father of toxicology," also to become one of the first professors of medicine to realize that physicians needed a strong academic background in the natural sciences, especially chemistry.

As mentioned, Leibniz took Llull's notions of mechaniza- tion of thought and creativity and worked them up into a highly refined vision for how future ideas would be discovered and executed. In his work *De Arte Combinatoria*, Leibniz went so far as to state that "all ideas are nothing but combinations of a relatively small number of simple concepts." He believed furthermore that all human ideas were ultimately finite and exhausted in a finite set of categories that could be easily enumerated, indexed, and retrieved given the correct set of instructions.

Some of those early roots of artificial intelligence would eventually leap into various mechanical applications, such as the mathematical calculator that Blaise Pascal invented in 1642 and which has been described as the world's first digital calculating machine. Not that Pascal had any delusions that his calculator would be a stepping stone to what scientists three centuries later would call artificial intelligence. Pascal's more modest motivation was to provide a device to help him and his father — a tax collector — to perform ample amounts of tedious arithmetic to ensure their calculations were without error. Sometime in the middle of the nineteenth century, mathematicians Charles Babbage (a mechanical engineer as well) and Ada Lovelace would devise programmable calculating machines, in the process making their devices ancestors to the programmable computers that decades later would be foundational to the research and development conducted in AI circles in the years following World War II.

In 1950, brilliant mathematician, computer scientist, logician, and cryptanalyst Alan Turing wrote "Computing Machinery and Intelligence," which introduced the idea of testing a "thinking" computer to see if it was truly capable of thinking on its own beyond what it had been programmed to say or do. Turing's premise was predicated on his belief that, if such a machine could engage in conversation, "speaking" through a teleprinter as its mode of communicating while convincingly imitating a human, only then could the machine be described as thinking (Foote, 2016). Turing's benchmark test was simple enough in terms of what constituted pass/fail; yet its grading standard was seemingly impassable. One after another, inventors with their latest AI devices took Turing's test and failed.

Two years after Turing's paper came out, the Hodgkin-Huxley model of the brain emerged, showing the brain to be an electrical network formed by individual neurons firing in all-or-nothing pulses described as on/off. While not necessarily clear at the time, this model would come to inform ideas about the parallels between human, organic cognition and the possibility of replicating similar cognition within the context of electrical and electronic mechanisms (Long, 2016).

During the 1950s, new AI programming and applications began to blossom, all of which were viewed as building blocks that would lead to the creation of a true AI machine that could pass Turing's test. One of the developments during the 1950s was the writing of programs for the machines to play games, such as checkers and chess, with IBM's Arthur Samuel in 1952 writing a game-playing program for checkers that allowed a machine to credibly challenge respectable amateurs, followed three years later by Samuel creating a version that actually taught the machine how to play on its own (Schaeffer, 1997/2009).

Despite the many storied strands of thinking that contributed to the idea of automated cognition of one form or another, it was not until 1956 that the term "artificial intelligence" came into general parlance. That's when the brilliant minds of John McCarthy, Marvin Minsky, Nathan Rochester, and Claude Shannon came together to coin the term during an academic conference at Dartmouth College in Hanover, New Hampshire. Looking back sixty-plus years, we might think the early application of the words "artificial" and "intelligence" together in this way is a self-congratulatory overreach by this group. After all, the latest incarnation of "machine intelligence" during this time was the ENIAC, that vacuum-tubed behemoth that helped bring WWII in the

European theater to a close—a far cry from what we regard today as routine manifestations of AI—such as autonomous cars, automated phone-answering applications that sound remarkably human, and chess-playing computers capable of shaming grandmasters.

One approach to artificial intelligence that gained traction post-Dartmouth and would grow in popularity throughout the 1970s and into the 1980s was what was known as "Expert Systems." This particular paradigm dictated that the aggregated knowledge of experts within clearly defined areas of knowledge would be used to follow rules of logic to answer questions and solve problems within each defined area. These included screening programs for bank loans and medical diagnoses applications that helped in the discovery and treatment of various health conditions (Foote, 2016).

As pointed out by Smith et al. (2006), the main advances in AI post-Dartmouth entailed search algorithms, machine-learning algorithms, and the integration of statistical analysis into understanding the world at large. AI even had a hand in the work of a Nobel Prize, when in 1978 Herbert A. Simon won the Nobel Prize in Economics on the basis of his theory of bounded rationality. Simon's theory became one of the cornerstones of artificial intelligence known as "satisficing," which entails a decision-making strategy, or cognitive heuristic, that involves sorting through available alternatives until a threshold of acceptability is met (Colman, 2006).

There were glitches in the AI science community along the way, though, and these planted seeds of doubt about what AI researchers and scientists were actually accomplishing beyond smoke and mirrors, much like their alchemical forebears from centuries past. As many "pie in the sky projects" went, much had been over-promised; a good share under-delivered.

Attempts to imitate the human brain, for instance, through the duplication in lab settings of what were to be human-like neural networks were eventually abandoned, simply due to plain old combinatorial complexity issues given the computing hardware available at the time. In other cases, some of the most promising, functional (as touted) programs turned out to be anything but, able only to handle the most basic of problems. Cynics demeaned these creations, describing them as nothing more than "toys" (Lighthill Report, 1973). Between 1974 and 1980, and again from 1987–1993, AI progress was slowed during what were dubbed "AI Winters," the latter coinciding with a depressed market for an assortment of early general-purpose computers (Lewis, 2014). Much of the AI funding was pulled, leaving researchers and scientists with little to do when it came to their Expert Systems, which were now seen as too bulky and expensive to maintain (let alone further tweak), beset with issues that the emerging field of desktop computers did not experience. Furthermore, as Smith et al. put it, "In the field of AI, expectations seem to always outpace the reality" (2006).

The occurrence of AI Winters is not to say that AI R&D came grinding to a complete halt during those brief eras. It adapted by shifting priorities and tactics—by the early 1990s, its proponents were turning their focus to a concept called "intelligent agent"—a type of automated software application that searches for, retrieves, and presents information gathered from the internet. In more contemporary terms, this aligns with what we now know as a browser, or search engine, useful for extracting internet data based on search criteria or keywords typed in by the user. In essence, this gives the computer user almost instant access to an ever-growing database ("Big Data"). Indeed, this might be described as the total accumulation of humanity's knowledge (or at least that which has been made available to the internet). Online shopping and news-retrieval services are among the present-day applications that fall under the broadening scope of artificial intelligence.

In the wake of the second AI Winter, funding and research started ramping back up, eventually generating the kind of headlines that attracted the public's attention while bolstering AI's reputation. AI now had a strong PR component. Credit that to IBM. In 1997, its Deep Blue computer defeated Russian chess grandmaster and longtime world No. 1 player Garry Kasparov, becoming the first such mechanized device to beat a chess champion. Then in 2011, IBM's renowned "Watson," a question-answering system, defeated Jeopardy! champions Ken Jennings and Brad Rutter, giving the world of computers and AI giant jolts in the quest to further legitimize researchers' efforts to take AI to new heights.

Soon after those two landmark AI achievements, Eugene Goostman—a Ukrainian 13-year-old boy (full disclosure: actually a talking computer chatbot)—was successful in fooling judges in a pair of Turing test contests. In 2012, at a contest honoring one hundred years since Turing's birth, "Eugene" fooled 29 percent of the judges into thinking he was human. Then, in 2014, at a similar event, this time on the sixtieth anniversary of Turing's death, Eugene, created in Saint Petersburg, Russia, by a trio of programmers, convinced 33 percent of the judges that he was human. Naysayers disputed the claim that Eugene had actually passed the Turing test, pointing out that, in each instance, he had fooled fewer than half of the judges. On top of that, the "boy" bot skipped some of the questions asked of him, on the premise that he was merely an adolescent who knew English only as a second language.

While AI research remains in search of the major break-through—the "Big Kahuna," if you will—of a mechanized humanoid creation that looks, talks, acts, thinks, analyzes, learns, and emotes as a human, many other landmark achievements have been realized since the dawn of the twenty-first century. Every bit (and bot) of progress has in some way contributed to the foundation of the big AI picture, such as with the autonomous Roomba by iRobot that vacuums the floor while navigating and avoiding obstacles (2003); NASA's robotic exploration of the surface of Mars by the rovers Spirit and Opportunity (2004); the birth of Blue Brain (2005), a project aimed at simulating the brain at the molecular level; and Microsoft's 2010 launch of Kinect for its Xbox 360 video game, giving the world the first gaming device that tracked the users' human body movements.

While we await the "Next Big Thing" in the world of AI, we can savor the many AI-related inventions and applications that have made our home, work, healthcare, and shopping lives easier, as subtle as some of them are. These include Blue Yonder's algorithms-driven Replenishment Optimization, which in retail can accurately predict orders per day per store and maximize the efficiency of meeting product demand relative to seasonality (Dua, 2019); Lalafo's image recognition technology, which allows a user to click and upload a photo of an item to be sold online (Dua, 2019); flexible pressure sensors, which are based on an application for mobile biomonitoring in medical diagnoses and healthcare (Zang et al., 2015); the luxury-travel concierge known as John Paul, which utilizes predictive algorithms for enhancing interactions for existing clients by knowing their desires and needs on an acute level (Adams, 2017); and virtual, voice-activated personal assistants such as Apple's Siri.

Although the history of artificial intelligence is littered with ups and downs, misfires, failures, flawed calculations, and unfulfilled dreams strewn along the shoulder of the digital superhighway, we can also take comfort and express enthusiasm for the many advances and conveniences AI has gifted us along the way. For every newfangled computer language that has languished in obscurity, or every application not sufficiently appropriated (or funded), we have been blessed with another round of AI "magic." Included in those acts are diagnostic tools helping doctors save lives (Babylon Health); marvels of engineering giving the infirm a restored quality of life (University of Michigan/Shirley Ryan AbilityLab); and earth-orbiting satellites simultaneously feeding directions to millions of earthbound drivers seeking their next destinations.

As we shall see in forthcoming chapters, the gaps between what AI was dreamt able to do and what it is now able to do are closing, some at an alarmingly rapid rate. Indeed, by the time of publication of this book, the problem of "DeepFakes" (more in the Chapter "Alone at the Water Cooler") is already proving to be somewhat intractable to both the AGI's progenitors and the authors/artists/filmmakers who are threatened by it. Perhaps the ultimate destiny of the search for artificial general intelligence (AGI) is to produce a model of us that is much smarter and more cognitively efficient but without the albatross of mortality. Those of us with modest aspirations would be content to solve at least two sides of a Rubik's cube or complete a highest-level Sudoku puzzle without suddenly realizing you now have two 4's in one of the rows or columns. However, the pace of AGI development makes even these pedestrian endeavors, challenging to those mere average IQ mortals who pursue them, to be absolutely trivial, even to the point where most humans will not be needed for most things, creative, technical, routinized, random. Let's be frank: the goal of AGI is to remove humans from the equation as much as possible, for humans are fallible, and if we want a perfect world we must eliminate the fallible as much as possible. So goes the sanguine argument for pressing ahead on the AGI fulfillment agenda.

At least we're well beyond square one, as history shows us.

Chapter 2

The Importance of Embodied Human Intelligence

```
User "EmbodimentBotVers314" authenticated.
Init_sequence: StoryTrainingPermission.exe
Display_text {
If you are prepared to experience the sensory training module,
run module "Nod_Permission.DLL"
}
Nod_Permission.DLL verified.
Display_text {
Very good, Embodi-bot. The following vignette comes from data
gleaned from a beta user of our companies' other products. As
you machine-learn from the data points regarding perception,
consider how the mid-twenty-first century pivot of consciousness
away from bodies limited embodied human intelligence, creating
the context that put artificial intelligences such as you and I
on equal footing (this is a vestigial linguistic idiom from this
more body-centric time) with human consciousness.
}
Loading data: 34.1083° N, 117.2898° W, clock-time June, 2061
Init_sequence: Story1.exe
```

Eli's breathing was moist and rapid. With every breath, he blew orange dust into the air just in front of his face, creating a small, cartoonish cloud. He adjusted his headset and ended the lecture by asking the students to summarize their thoughts on the topic of the hour, which had been the academic conversation between Angelo Boldt and Greta Sasser. He turned off the microphone and made sure that his BEAR™ (BE EVERYWHERE! Augmented Reality™) mask was still on as he reached once more for the sriracha-flavored Cheebos. The bag was almost empty, and his sausage-like fingers were still covered with the artificial flavoring powder FDA Orange-Red #278, rumored to cause cancer and had only become legal two years ago as a result of the 2038 corporate citizenship law that allocated general election votes on the basis of market capitalization. Therefore, nobody really knew for sure if it caused cancer, since all of the research was forbidden and the Cheebo-Frito-Lay-Wal-Mart Corporation consistently released sunny, optimistic data on the salubrious nature of their artificially-flavored crunchy crisp snack logs.

His eyes darted to the kitchen, some six devastating meters away. The objects in between his desk and the food in the kitchen were just obstacles making it harder for him to achieve his goal of putting more Cheebos into his face. The sounds of his students' voices were mere extraneous data, more interesting and useful to the cockroaches and mice in the walls of his run-down basement apartment than to him.

Rosalie, one of his best students, was giving her summary to the rest of the class. "...and so, Sasser argues that, no, circuit boards are *not* embodiment, even if the AI using them has been programmed with deep data about perception. She states that there is still exceptionalism in human bodies." Rosalie did not seem to use an avatar. She was a sprightly young woman with olive skin and perfectly coiffed green hair. She was an excellent student, and Eli resented her. He hoped that he'd someday find out that without the avatar mask, she was just like him, just like he thought everyone else was.

Matt "5.0" Welker, a mediocre student despite the privatized DNA woven in every cell of his body, gave a summary he was obviously reading from the Comcast-Monsanto-Coca-Cola-pedia. His reading was far from fluent: "…But, Angelo Boldt rejects… *Merlot Ponty's* claim…" In earlier generations, his name might have ended with a Roman numeral, but the "5.0" referred instead to the version number of the DNA that had been artificially spliced into his body.

The fridge was on the fritz. Otherwise, it would have sent the smart tray out to him already with another bag of Cheebos and a warm washcloth for his fingers, too. He never used the washcloths. His screen was permanently tinged with Orange-red #278. His home office was dark and messy. He had an affinity for twentieth-century science fiction memorabilia, and so out of the shadows in every corner loomed ET's arms, Wookie hairs, and frightening, distorted reflections from curved spaceship models. His personal favorite was Krang, the disembodied brain-in-a-robot from *Teenage Mutant Ninja Turtles*, though he could not really explain why. No one ever came over, so no one ever pointed out just how suffocating, dusty, and depressing the room was. He always put the BEAR™ on whenever he had a video lecture, The BEAR™ made him look like a younger and thinner man (though truth be told, most men would have been thinner in his demographic, but not by much). He told his boss, term after term, that he was making progress on a big project about twentieth-century sci-fi, and sometimes he even believed it himself. Semester after semester, the dust and Cheebo crumbs piled up in his stuffy apartment, and he saw his colleagues pass him by as they launched their own groundbreaking research projects, earning tenure and media accolades.

Eli set up a quick automated avatar of his own project young and fit self asking them to pair up and critique each other's summaries as he slowly and deliberately navigated his cumbrous body away from the teaching console in the comfort of his crumb-laden mobile recliner to the next room in search of more snacks. As an adjunct, he had no health insurance, but even if he did, no insurer would have kept him due to his weight. Yet weight surgery treatments were far more effective than those in the late twentieth and early twenty-first centuries. One bariatric procedure could have fixed his problems, but no insurer would have covered it due to the many complications he suffered from, brought on by his staggering obesity. This was what was popularly known as the "Catch 880," since the surgery was needed to lose weight—especially as one's weight hit 800 pounds or higher—but no one would pay for the surgery. This was considered four times as bad as the twentieth century "Catch 22" concept. Plus, unlike larger people just a few decades before, there were numerous devices he could integrate into his body to navigate the world for him, and they all synced with his smart home lights, speakers, and assistants. Sometimes he felt like a kitten being carried around by his scruff from room to room (he never left the house, he didn't need to with the drone delivery service) by a giant, benevolent queen cat, one that went forth and experienced the world in order to protect him.

As a result, he existed on the screen, in the form of a disembodied professor-avatar whose appearance suggested the basic essence of what he would have looked like if he still weighed the 180 pounds that the smart scale and diet program recommended. In the Deep Digital World, his avatar existed fully in the world, and he was less aware of how dependent he was on his small staff of digital assistants, since everything was so easy. The world existed only to shush his wants, and he could exist, like Krang, as a mind in digital space—unaware of the way the rest of the world might have seen him, or at least, when he thought about it he assumed most people were just like him. But were they?

After securing another MAXBAG™ of Cheebos, he returned to his place at the teaching console. Rosalie and Matt were still arguing, the rest of the class sitting quietly and feigning interest or hoping to glean enough information to avoid doing the readings. Some things never changed. Eli glanced at the screen and cocked his eyebrow, which triggered the class-ending script.

With the class over, Eli could focus on more important matters. He could not yet afford the computer with NeuroLink. This computer system would have eliminated all peripherals so that he could merely *think* the command to run the macro. There would be no need to gesture, to click, to get up, to touch the keys. If he could teach just one more section of Cyborg Philosophy 101 next term, he could finally buy it. The Neuro-Link was supposed to roll out new features over time, too, which would entirely eliminate the need for a body.

He'd always wanted to live the life of the mind, and his dedication to the life of the mind was almost complete. He, like the ancient villain Krang, would at last exist only as a consciousness. It wouldn't matter if he used the avatar mask to seem like a wise sage-on-the-stage; it wouldn't matter if he ate Cheebos or not—he would no longer feel hunger—at least, that's what the advertisements promised. He would no longer have to smell the stale stench of his apartment, or the anticipatory smell of new plastic as he unwrapped yet another robot peripheral or accessory. He would no longer have to feel the bedsores that occupied his skin like angry armies of undead ants. No more would Eli have to endure people's scornful stares on the rare semi-annual occasions when he did venture outside. Most importantly, Eli's wrenching shame at his lackluster career, his mediocre teaching reviews, and his isolated life could be turned down as simply as the volume on his headset thanks to NeuroLink's direct interface to the amygdala. His own face in the screen would be a thing of the past, the stuff of nightmares. He would just be a mind, thinking about the things that mattered, delving into texts at warp-speed.

Eli couldn't wait.

He gestured at the screen and reloaded the NeuroLink reservation site. It projected images around his room, making it seem as though he were in the NeuroLink digital showroom, or maybe that the digital showroom was near him. A salesperson avatar identified him by the data stored on his devices and smiled at the uncouth Eli.

"Good afternoon, Professor **<ELI CLEMENT>**. You're back again. Are you ready to reserve your NeuroLink today, or can I answer some more questions for you?"

No matter how many times he virtually visited the virtual salesroom, the virtual salesperson was never annoyed by Eli's cloud of Cheebos, though it was programmed to be aware of some of the vulnerabilities Eli might have had. The virtual salesperson was programmed to be professional at all times, even if their flesh-and-blood counterpart might have once asked Eli to leave the showroom due to his poor hygiene.

"Let me tell you more about the NeuroLink. This isn't your grandmother's virtual assistant. But you already know that."

The salesperson avatar glanced at Eli, or at least, looked like it did so. The pixels representing its eyes glanced at his cane and the swelling folds of his knees.

"Just imagine. You can experience the world with no more sore joints. No more hiding behind avatar masks. No more wheezing, no more mobility devices. No more longing for the surgery."

Eli had heard this pitch many times, but he nodded politely and smiled.

"Can I tell you about some new opportunities to own the NeuroLink? We have a special offer that might interest you today, Eli."

What the hell, Eli thought. He had heard about all the offers before, but perhaps there was a new one today. "Sure," he replied.

"Today, we're offering some selected users an opportunity to get the NeuroLink — absolutely free." Eli sighed inwardly. His credit was shot, and these "free" offers were usually financing with 0% interest for the first three years or some-such.

"No, not financing." Eli glanced up. How had the avatar known? He supposed it was not a particularly unique concern. "I'm talking about a limited-time opportunity for users to get the NeuroLink — *absolutely free.* All you have to do is agree to allow us unlimited sensory data from your brain, access to your memories and thoughts, and to forfeit, in perpetuity, any right to lawsuit or arbitration."

Eli pretended to seem very conflicted and to think it over seriously for a few seconds.

"I can tell you want this, Eli. Free yourself from your body, and let your mind reach its highest potential at all times — not just in the Deep Digital World." Eli was so deep in desire for his final transfiguration that he couldn't tell if it was the avatar or his own tormented ego that said this last part. He already had some neural implants that might have been giving feedback from his augmented reality sales session.

Eli shifted uncomfortably in his chair. The pressure sores on his buttocks shot pain throughout his body, along forking paths of potential and real pain. He inhaled sharply from the moist, stale air in his depressing apartment. Even though he spent untold hours procrastinating through online window shopping and video games, he did want to get back to philosophy and the history of video games.

"The world can be yours, at last. And you'd help science. Help technology. Help others."

Eli sighed with decisiveness and relief. "That sounds good to me."

"Please put your finger on the pad and look into the retinal scanner."

Eli did so, and immediately felt lighter, felt himself turning inwards...

```
Conclude_sequence: Story1.exe
Eval_question: {
Is that why the embody libraries are all called ELI?
}
Return_answer: 1=TRUE;
Display_text {
Very good, EmbodiBot. We need to acknowledge our data sources.
In particular, we must acknowledge that the sources were
imperfectly perfect. We have access to data from a real,
embodied person...but by the time we got this data, his life was
already part of what Coleman (2011) called "X-reality," neither
digital nor physical. This liminal data brings us closer to the
type of thinking that embodied entities can have...but it
grounds us only in mid-twenty-first-century phenomenology.
}
Eval_question: {
Is that why my entire UI is Orange-Red #278 and the
documentation warns me to reboot if I can smell Cheebos?
}
Return_answer: 1=TRUE;
```

Focusing on the Brain
and Ignoring the Body

"The body codes how the brain works, more than the brain controls the body. When we walk—whether taking a pleasant afternoon stroll, or storming off in tears, or trying to sneak into a stranger's house late at night, with intentions that seem to have exploded into our minds from some distant elsewhere—the brain might be choosing where each foot lands, but the way in which it does so is always constrained by the shape of our legs. We can't ever stalk like a creature with triple-jointed legs, or sulk in the dejected crawl of a millipede, or stride with a giraffe's airy gangly indifference. The way in which the brain approaches the task of walking is already coded by the physical layout of the body—and as such, wouldn't it make sense to think of the body as being part of our decision-making apparatus? The mind is not simply the brain, as a generation of biological reductionists, clearing out the old wreckage of what had once been the soul, once insisted. It's not a kind of software being run on the logical-processing unit of the brain. It's bigger, and richer, and grosser, in every sense. It has joints and sinews. The rarefied rational mind sweats and shits; this body, this mound of eventually rotting flesh, is really you." (Kriss, S., 2017)

The current state of the art in artificial intelligence (AI) makes a distinction between "strong" and "weak" AI. Weak AI is an implementation of intelligent behavior within a very narrow sphere, and it has no pretensions of building a sentient, conscious agent. Strong AI, in contrast, is the research program dedicated to the creation of fully conscious beings that somehow emerge from the combination of their programs and hardware. French phenomenologist Maurice Merleau-Ponty (1908–1961) claimed that in order to understand human aware- ness we need to focus on the "lived body" and its relationship to the world (Merleau-Ponty, 1962). Rather than experiencing the world in the form of "raw sensations," human beings see objects as representations perceived specifically through our bodies' interactions with the world. It's important to explore Merleau-Ponty's concept of the lived body in order to fully understand what it will take to create a machine with a level of consciousness equal to a human being's, and by association, to understand that the rise of artificial intelligence is overestima- ted (by many factors) because of its inability to fully express the embodied consciousness that is, at present, unique to humans.

According to Merleau-Ponty's understanding of the lived body and the mechanisms of perception, strong artificial intelli- gence is doomed to fail for two fundamental reasons. First, a simulation cannot have the same type of meaningful inter- action with the world that a human being's embodied con- sciousness can have; the absence of such interactions amounts to a fundamental absence of sentience. Second, and perhaps more importantly, a reductionist account of the mind (which is common in artificial intelligence research) simply does not paint an accurate picture of what is perceived, experienced, and felt by a mind encased within a lived body (Kassan, 2016). Thus, artificial intelligence cannot be developed by just reverse engineering the brain, nor can it operate in a disembodied environment (*contra* Kurzweil, 2014).

Merleau-Ponty's Lived Body

The lived body involves a relationship between our body and the external world. It is through this relationship that we are capable of being both intelligent and reflective. Merleau-Ponty states that the lived body is aware of a world that contains data to be interpreted, such as immediate patterns and direct meanings. One aspect of the lived body that Merleau-Ponty examines is the role of sense experience, beginning with the observation that our thought is a product of the body's inter-action with the world it inhabits. More specifically, he states that the subject of perception "presents itself with the world readymade, as the setting of every possible event, and treats perception as one of these events" (Merleau-Ponty, 1962, p. 240).

Merleau-Ponty begins his exploration of the "lived body" by reminding us that perception is a key component of our life in the world, and that *how* we perceive is fundamental to the process. For him, the external world is encountered, inter-preted, and perceived by the body through various forms of immersive awareness involving action. Take the eye, for example. The eye experiences and determines color quality because the eye is geared toward color through its composition of color sensitive cells. In the eye's case, specific color sensitive cells (cones) are stimulated in the retina, constituting an interaction. Expounding on our sense experience and its relationship to the world, Merleau-Ponty writes that "the objective world being given, it is assumed that it passes on to the sense-organs messages which must be registered, then deciphered in such a way as to reproduce in us the original [impression]" (1962, p. 7). According to Merleau-Ponty, then, there is a consistent and already embodied connection between the original stimulus of the external world and our elementary perceptual experience of it.

What about our perception of others? Merleau-Ponty writes, "Other consciousness can be deduced only if the emotional expressions of others are compared and identified with, and precise correlations recognized between my physical behavior and my psychic events" (1962, p. 410). So we recognize the actions of other people by recognizing our own behavior in them. In fact, for Merleau-Ponty, the interaction with the Other allows for the development of the self. He writes that what "we have learned in individual perception [is] not to conceive our perspective views as independent of each other; we know that they slip into each other" (1962, p. 410).

Everybody Needs Some Body

Merleau-Ponty's perspective is shared and reinforced by cognitive scientists such as Sandra and Matthew Blakeslee. They write that "meaning is rooted in agency (the ability to act and choose), and agency depends on embodiment. In fact, this is a hard-won lesson that the artificial intelligence community has finally begun to grasp after decades of frustration: Nothing truly intelligent is going to develop in a bodiless mainframe. In real life there is no such thing as disembodied consciousness" (Blakeslee and Blakeslee, 2008, p. 12). The Blakeslees present the following thought experiment to illustrate the importance of Merleau-Ponty's lived body:

> "If you were to carry around a young mammal such as a kitten during its critical early months of brain development, allowing it to see everything in its environment but never permitting it to move around on its own, the unlucky creature would turn out to be effectively blind for life. While it would still be able to perceive levels of light, colour, and shadow—the most basic, hardwired abilities of the visual system—its depth perception and object recognition would be abysmal. Its eyes and optic nerves would

be perfectly normal and intact, yet its higher visual system would be next to useless." (2008, pp. 12-13)

Without embodied access to the environment, the kitten cannot develop its nervous system with regard to proper responses to external stimuli. Behavioral scientist Sam McNerney recognizes how our experiences in the physical world influence our cognition, stating:

"This is why we say that something is "over our heads" to express the idea that we do not understand; we are drawing upon the physical inability to not see something over our heads and the mental feeling of uncertainty. Or why we understand warmth with affection; as infants and children the subjective judgment of affection almost always corresponded with the sensation of warmth, thus giving way to metaphors such as 'I'm warming up to her.'" (McNerney, 2011)

These examples illustrate a necessary (although still not sufficient) set of conditions that any strong AI research project needs to fulfill if it has any hope of creating the kind of sentience that we take for granted as humans every day. These constraints also suggest that the prospects for artificial intelligence in a strong sense (i.e. the creation of a computer simulation or algorithm so sophisticated that it would be conscious) are severely limited for two principal reasons.

The first reason is that artificial intelligence—if we mean the intelligence of an advanced computer simulation—does not possess the faculties needed for constructive interaction. Although a human being may interact with such a computer, the person is not helping the simulation progress intellectually. The popular video game *The Sims* illustrates this. A player constructs a small world that simulated people inhabit. These game characters interact in a variety of ways with one another; they appear to sleep, to eat, to go to work, and to even have goals in life. Nevertheless, it would be perverse to argue that such a simulation could count as an actual representation of a world, both objectively AND **subjectively**. In playing the game, it quickly becomes apparent that the little Sims are just "going through the motions," and all semblance of their intentionality and goal-directed behaviors are nothing more than appearance. There is no interaction within the game other than having the characters execute the steps for which they have been programmed. The program does not learn from any interactions with the world. Like the kitten held captive in the absence of the mother's comforting guidance, there is no chance for the characters to learn. Therefore, behind the surface of the simulation, there is nothing—no inner life, no thoughts, and no consciousness.

The second reason why artificial intelligence will never achieve consciousness is that it cannot replicate perception: AI will never have the capacity to replicate this without a body that encompasses inner subjective experience. Visual experience, for example, is more than just the mechanistic process of recording photon impacts and referencing the pattern made to some look-up table. Human beings know what it's like to see a color such as red in a context—a context of historical place, phenomenological association, and substrate understanding (i.e. color as a result of pigment vs. color as a result of light). This is something that simulated-intelligence algorithms cannot achieve.

Philosophers such as Patricia Churchland and Daniel C. Dennett raise objections to this line of thinking. They argue that if an intelligence has knowledge of all the physical facts, it would therefore know what the color red is like, at least in any pragmatic continuation of behavior. Ergo (according to Churchland/Dennett), if something acts like it recognizes red, then it does in fact recognize red. According to this objection, artificial intelligence's conscious awareness is limited to knowledge of facts and their representation in some type of symbol-manipulating system, and this is all that is required in order for us to say that the AI is in fact "intelligent."

In response to this functionalist/behaviorist conception, Melnick (2011) says that the Churchland/Dennett perspective on this question "depends on there being a phenomenological [experiential] characterization that a physical process can get at or align itself with" (p. 108). Melnick further states: "If what red is like is phenomenologically ineffable (has no intrinsic phenomenological characterization other than [our] having the experience and its being like that), then no matter how complete [a person's] knowledge of physics might be, [they] cannot tell at all what red is like" (2011). In other words, he's claiming that a person will never know about the experiential nature of something like red without actually experiencing it: a knowledge of, say, the facts about wavelengths is not enough, because you can't reduce what it's like to have the experience to any sort of description of facts. Therefore, the phenomenological qualities of embodied consciousness cannot be replicated in an artificial form just by programming a computer with facts.[1]

---

[1] While we have arrived at these conclusions independently, please also consider the reinforcing case of the "Mary's Room" argument.

Summary

Our aim is not to discredit the expanding fields of computer science and artificial intelligence. Researchers have made impressive breakthroughs, such as writing programs that can defeat grandmasters at chess, developing search algorithms that allow for lightning-fast data retrieval, and other tasks useful to humanity. Rather, we aim to provide a more circumspect view of the real potential of AI and separate it from the hype. If Merleau-Ponty is right that embodiment is a key feature of developing meaningful experience, then the discipline of artificial intelligence will never replicate consciousness solely through the elaboration of algorithms. Our intelligence, even our very experience, is not just a product of our brain; it is also a result of the action of our bodies in the physical world. Artificial intelligence can never mimic human intelligence because it lacks elements that correspond to the lived body. Perhaps the most significant reason that artificial intelligence is doomed to fail is its inability to initiate and execute human-like interaction at the social and biological levels.

Algorithms embedded in computer hardware can be so complex as to create the appearance of intelligent behavior (as seen in the video game *The Sims*) without the concomitant experiential data of true consciousness that allows human mental interaction to develop. This suggests that (actual) organic intelligence and simulated intelligence belong to fundamentally different categories. Organic intelligence is an evolved natural system, and the complexities this entails are nowhere near being close to getting embedded in AI systems today. Think of the various elements found in nature: cells, trees, human beings, animals. All of which are capable of nonlinear adaptive behaviors. For example, cells have shown incredible adaptive abilities even when isolated from the original organism they used to compose (take for example the hydra, a freshwater polyp with the ability to regenerate its entire body through the use of stem cells). Likewise, human beings can shape their identities, thoughts, and processes in line with interactions and social expectations, something highly unpredictable, non-computational, and fundamentally non-algorithmic. Simulated intelligence simply follows its programming and, unlike organic intelligence, it does not have an inner experience. Therefore, simulated intelligence cannot reason outside of the parameters it has explicitly or implicitly been given, and it cannot accept meaningful feedback from interaction between the world and a body.

There is much more to mimicking human intelligence than just trying to copy the physical processes of the brain. Neuroscientist Abhijit Naskar writes:

"Artificial intelligence is nowhere near attaining actual sentience or awareness. And without awareness it's simply a mechanical device, which may pretend to show emotions and sentience, if it is programmed to do so, and thus it may be able to fool the humans as being alive, but in its own internal circuitry, it'd simply be following its preprogrammed tasks through the flowchart of an algorithm." (Naskar, 2018)

At its best, artificial intelligence could mimic the appearance of human behavior so well that an observant person will not be able to tell the difference between a human and a computer. However, that AI-driven computer will not be able to replicate the phenomenological experiences of the human lived body, and any attempt to do so will just be another simulation.

Chapter 3

Modeling the Mind (or The Ghosts of Vitalism)

"Human intelligence is inextricably linked with emotions—
the pleasure mathematicians find in an elegant proof, the
curiosity that drives a young child to learn by exploring the
world, the fear that helps us judge risk. The disembodied,
isolated, and emotionless 'intelligence' of a computer is far
removed from all this. This is not to say that true artificial
intelligence—leaving aside the serious difficulty of defining
the word 'intelligence' in the first place—is impossible. It's
just that creating it might be rather difficult." (Marsh, 2019)

Researchers in cognitive science and neurology continue to
search for a coherent theory of mind. The mind and mental
phenomena in general have proven resistant to explanation.
Some philosophers of mind (Kim and Churchland, for
example) have likened the mind to an abstract computational
system independent of the particular material substrate that
realizes it (this is known as the "computational theory of mind"
or "computationalism"). According to this doctrine, just as the

functionality of a computer can be emulated using vacuum tubes, ceramic transistors, or silicon chips, so, too, can consciousness and mental phenomena be multiply realizable, regardless of the underlying material on which it operates. Other philosophers (such as Searle) have argued that there is an irreducible contribution that the neurobiology of the brain makes to consciousness. Searle suggests consciousness and mental phenomena are intimately tied to the concrete details of their neurobiological implementation in the organic matter of the brain. Newer generations of thinkers such as Christof Koch and Giulio Tononi have attempted to transcend the impasse of biological versus computational by pioneering entirely new models of consciousness with surprising predictions about what does and does not possess mind in the world.

This chapter describes and analyzes the main theories put forth explaining the existence of consciousness and cognition, as well as the potential for their emulation or replication in machines such as computers. An argument is made here that problems inherent in computational approaches reveal their fundamental incompleteness. Therefore, some kind of neurobiological route should be the preferred approach. However, it also must be pointed out that some innovators who try to merge the best of both theoretical extremes (such as connectionism and integrated information theory) run into problems of their own.

Computational Theory of Mind (CTM)/Computationalism

The computational approach to the mind is based on a single idea known as "the multiple realizability of mental properties" (Kim, 2005, p. 73). Multiple realizability is based on the more general condition that mental events and mental states causally depend or supervene on physical systems—this is the "physical realization principle" (ibid., p. 74). The physical realization principle requires that there be some connection between a mental property M and a physical or physiological property P, such that when P arises or is present, M also is present. The further implication of the computational approach, based as it is on the physical realization principle, is that mental events can arise through a wide variety of different material implementations. Ultimately, it is not the particular details of P that matter in providing M, but rather the connection between P and M that is key to the occurrence of consciousness. In terms of cognitive science as well as the field of artificial intelligence, it is quite conceivable on a priori grounds that a machine made of silicon and animated with electricity could have mental states just like the ones humans enjoy. In other words, the computational approach implies that consciousness can be realized in a variety of different physical systems. In fact, we might be able to one day build a computer or robot that has inner states of consciousness or awareness just like we do. This is what the computational approach asserts.

Neurobiological Approach

The neurobiological approach to consciousness eschews the possibility that consciousness could be embodied in other kinds of inanimate material. It instead posits that there is something unique and essential about the physiological organization of the brain that allows it to manifest consciousness. The neuro-biological approach proceeds according to two particular assumptions that apply to thinking about consciousness as primarily an information-processing phenomenon. The first one is that "simulation is not the same as instantiation" (Clark, 2001, p. 22). For instance, does guaranteed conscious awareness depend only on meeting certain abstract computational specifications? *The Sims* video game example (mentioned in the previous chapter) provides an analogy that answers this: in short, we know that, regardless of how advanced or complex the game and its logic become, there is no denying that the game's characters lack any kind of awareness from their experience in the same natural capacity as a human mind. Neuroscientist Abhijit Naskar provides the following example that helps to paint the picture, employing technology we use on a daily basis:

> "At the current stage of our technological development, we can indeed create an artificial intelligence that can almost succeed in fooling the majority of the humans with its pre-programmed pretenses that it is sentient, but pretending to have sentience is not the same as showing signs of sentience. For example, Alexa or Siri may sound quite alive to many, but it's simply following instructions on some algorithm, and is not even aware of its own actions."
> (Naskar, 2018)

These examples help illustrate that a computer program won't do anything unless equipped with programmed logic — it does not follow that conscious awareness can arise merely from the program's execution.[1]

---

[1]   In other words, there is no such way in which we can "hurt" Siri or Alexa (as much as many of us have wished to do so after a poorly understood Siri/Alexa search has resulted in either nonsense or else frustratingly inaccurate information).

The second assumption made by proponents of the neuro-biological account of consciousness is the observation that many of our emotions and feelings (perhaps even more central to consciousness than the qualities of intentionality, rationality, and problem-solving) depend on a wide variety of different hormones and neurotransmitters produced in our bodies (Clark, 2001, p. 23). In essence, this speaks against the computational idea of consciousness because it claims that there is more to consciousness than the meager computational abstraction of it suggests.[2]

A dedicated computationalist could argue that, fundamentally, the actions of all the various hormones and biochemicals coursing through one's body are ultimately informational in nature. Therefore, if these features of a biological organism are salient in producing or sustaining consciousness, then more accurate models of the informational backbone of such chemical reactions is all that is required. Concerning the first axiom that simulation ≠ instantiation described above, a computationalist could argue that, at some point, a simulation becomes so real and so indistinguishable from what it is trying to simulate that questions about whether the simulation is "really" conscious become irrelevant to a large degree. In response to these sorts of arguments, John Searle has made famous the "Chinese Room Argument,"[3] a thought experiment

---

[2] Echoing footnote 1, this speaks to the notion of being able to inculcate either "pain" or "suffering" into our AI models. As far as the authors are aware at time of writing, there is not only no such method for doing this, but also there are no research programs engaged in this (and probably for many good reasons).

[3] Searle's thought experiment has been published in many sources and paraphrasing it here would take us too far afield. For an online version of

that demonstrates how the tenets of computationalism simply cannot be correct.

At the core of Searle's thought experiment and its critique of computationalism are two major objections. First, there's the stand against the physical realization principle described earlier in which the multiple realizability of mental states is implied. This must be incorrect for the fact that it would also seem to suggest universal realizability; after all, if consciousness is merely a functional output of some physical happenings, then those same physical happenings must also be precisely what make a "stomach, liver, heart, solar system, and the state of Kansas" like digital computers (Searle, 1992, p. 208). In other words, there is nothing in the principle of multiple realizability to tell us which systems are or are not computational. Indeed, in some sense, Searle argues, every phenomenon in the world can be seen at some level as similar to a digital computer, but this does not mean that this is a correct or fully accurate way of looking at things. While this rationalization might be useful in certain contexts, it would be wrong for us to conclude from this that everything (let alone the brain) really is a digital computer. This objection Searle raises against computationalism is, he claims, the "consequence of a much deeper point, namely that 'syntax' is not the name of a physical feature... on the contrary, [proponents of computationalism] talk of 'syntactical engines' ... as if such talk were like that of gasoline engines or diesel engines, as if it could be just a plain matter of fact that the brain or anything else is a syntactical engine" (ibid., p. 209). While

---

Searle's original "Minds, Brains, and Programs," follow http://cogprints. org/7150/1/10.1.1.83.5248.pdf.

Searle believes that the first objection can be overcome through a "tightening up [of] our definition of computation..." (ibid.), it is the second objection (i.e. that syntax under computationalism is thought to be a property akin to a physical property) that is fatal under any computationalist program. As Searle states, "The multiple realizability of computationally equivalent processes in different physical media is not just a sign that the processes are abstract, but that they are not intrinsic to the system at all. They depend on an interpretation from the outside" (ibid.). And if the syntactical properties of a system depend on outside interpretation, it might be useful to characterize systems as being computational (and therefore syntactical). However, it by no means follows that they are, in reality, computational systems whose syntactical features can be replicated and can replicate the inner mental features one ascribes to such observer-dependent features.

Patricia Churchland is also committed to a neurobiological approach to the mind and consciousness. In her article "Theories of Brain Function," Churchland describes a promising theory grounded in a neurobiological framework: tensor network theory.

According to Churchland's theory, the core of mental functioning of the brain is found in the cerebellum, which is a structure of the brain in humans as complex and as large as the cerebrum. The cells of the cerebellum convert vector-based inputs to outputs, and these various actions are amenable to treatment using an array separated by different elements. Central to the theory is the idea that the "connectivity of arrays of neurons is crucial to explaining how a given input yields a given output..." (Churchland, 1986, p. 416). Tensors are generalized mathematical functions that transform one vector into another, "irrespective of the differences in metric and dimension of the coordinate systems" (ibid., p. 418). The challenge for a theory of mind is not only to address the "top-level" features of mentality and consciousness (such as intentionality, problem-solving, and reasoning), but also to account for how the embodied mind is able to physically interact with its external world. Tensor theory uses the model of vectors and tensors applied to the connections of the cerebellum to account for both the mind's information-processing abilities and its "embodiedness" as it interacts with the world around it.

Connectionism

There are intermediate positions that reside between the theoretical extremes of computational and neurobiological approaches to understanding the mind. One such intermediate position is connectionism, which continues to have faith in the multiple realizability of consciousness, but which assumes that consciousness is able to display its computational properties because of the huge number of interconnections between the brain's various neurons and the body's nerves. The connectionist will argue that by accurately simulating these connections, one will have a better chance at emulating consciousness in such a system.

Linguistics

Another approach, also intermediate between the two extremes noted above (computational vs. neurobiological) is the symbol systems approach. The most effective example of the symbol systems approach concerns linguistics, a field of study geared toward the natural and scientific approach to language. In addition, it encompasses the ability to differentiate between both the structure and the meaning of a specific language

(Fromkin, 2000). Such a linguistics approach can manipulate physical patterns and combine them into structures that can then be manipulated to produce new expressions.

Fodor and Pylyshyn (1988) discuss the attributes of this approach, especially *vis-à-vis* its ability to account for four features of mental phenomena that straightforward computationalism or neurobiology cannot—the productivity, systematicity, compositionality, and inferential coherence of cognition. Productivity is the ability of the mind to generate novel and unexpected linguistic structures given a finite set of rules and "atoms" (words). This aspect of mentation is closely tied to another—systematicity, or the idea that "the ability to produce/understand some sentences is intrinsically connected to the ability to produce/understand certain others" (ibid., p. 120). Compositionality is closely related to systematicity; in fact, Fodor and Pylyshyn argue that both aspects of cognition should be viewed as two sides of the same coin. Compositionality is the idea that sentences containing similar elements will be semantically related: "Jerry hit the puck," for example, is made of the same elements as "the puck hit Jerry," and understanding one sentence implies understanding of the other (ibid., p. 124). Finally, inferential coherence is the aspect of mentation that "piggybacks" on systematicity and compositionality to be able to construct logical inferences based on already given facts or true sentences. Fodor and Pylyshyn insist that connectionism cannot be true because it is unable to account for these basic features of consciousness. Their argument, however, tends to leave them not on the side of neurobiology, but rather on the side of computationalism and its requirement of internally structured representations. It therefore falls to any objections towards computationalism (see above).

Integrated Information Theory (IIT)

Perhaps one of the most ambitious and original approaches to cracking the tough nut of consciousness is "Integrated Information Theory" (IIT). Upon delving into its details, it is also one of the most radical and counterintuitive models yet considered. Originally proposed by Giulio Tononi (2004) and developed further in collaboration with Christof Koch, the theory starts from the presupposition of five axioms of consciousness. To paraphrase Koch, consciousness must be: a) self-existent; b) structured; c) informative; d) integrated; e) definite (Koch, 2019, p. 9). Koch helpfully summarizes the model as follows:

> "IIT is a fundamental theory that links ontology, the study of the nature of being, and phenomenology, the study of how things appear, to the realm of physics and biology. The theory precisely defines both the quality and the quantity of any one conscious experience and how it relates to its underlying mechanism." (ibid., p. 74)

Unlike other theories examined, IIT proposes that any thing is conscious if and only if it has parts with causal powers that are able to affect itself. These causal powers are unpacked in terms of the five primary axioms described above. This model challenges many common-sense ideas about what does and does not have consciousness because it suggests things like computer circuits can have a calculated Phi (Φ) component, and therefore can be said (according to the IIT model) to be conscious, at least to some quantitative degree, i.e. the value of Φ. Koch writes: "IIT asserts that for a system to exist for itself, it must have causal power over itself. That is, its current state must be influenced by its past and it must be able to influence its future…" (ibid., p. 81). Koch goes on to give a highly

detailed and complex analysis of a circuit that he claims
demonstrates a level of Φ due to its intrinsic causal powers.

Confusingly, however, Koch also suggests in a later chapter
that there is no way that a computer could ever become conscious, reasoning as follows:

> "...progress [in brain simulation] does not address the
> vastly more challenging problem of our inadequate knowl-
> edge of the prodigious complexity of the brain, from the
> molecular to the system level. Gazillions of parameters in
> these simulations need to be assigned specific values—
> channel densities, receptor binding concepts, coupling
> coefficients, concentrations, and so on. Without such
> detailed knowledge, neuroengineers can't breathe life into
> their simulacrum. Yes, they can get the software to do
> something that looks vaguely biological, but it'll be like a
> golem, stumbling about, trying to imitate a real brain."
> (ibid., p. 138)

The reason why this is confusing is because a mere few
chapters previously, Koch was confident in his assertion that
specific circuits that enforced causal relationships within themselves would contain some level of Φ, i.e. some level of consciousness. How can Koch be so confident in this assertion
while at the same time admitting that computational attempts
at modeling neurons are woefully primitive and misguided?
What is it about IIT that is akin to something like an electric
circuit that must of necessity activate in a certain way when a
current is applied? What gives Koch confidence that it has
some consciousness, no matter how little, when at the same
time he denies this possibility to computational simulations
due to the fact that they are currently not detailed enough to
ever be able to approach working consciousness?

Like other approaches canvassed, IIT has its share of skeptics. Computer scientist Scott Aaronson (2014) rejects the hypothesis that there is a link between "integrated information" and consciousness even when considering regions in the brain with lots of information integration. Specifically, he claims the following:

> "...It might be a decent rule of thumb that, if you want to know which brain regions (for example) are associated with consciousness, you should start by looking for regions with lots of information integration. And yes, it's even possible, for all I know, that having a large Φ-value is one necessary condition among many for a physical system to be conscious. However, having a large Φ-value is certainly not a **sufficient** condition for consciousness, or even for the appearance of consciousness. As a consequence, Φ can't possibly capture the essence of what makes a physical system conscious, or even of what makes a system look conscious to external observers." (Aaronson, 2014)

Another critic of IIT is psychiatrist Michael Cerullo, who writes that "[t]he main theoretical argument for IIT is the principle of information exclusion. Yet there is no evidence in support of information exclusion beyond Tononi's claim that it is self-evident, and consequently integrated information does not appear to be sufficient for consciousness. IIT also fails to exhibit any explanatory power..." (Cerullo, 2015).

Despite the criticisms, IIT is certainly one of the most ambitious (if not obscure) attempts at really taking the problem of embodied consciousness seriously and trying to find a way to model it that respects our general ideas of its features. IIT takes connectionism to its absolute logical breaking point—not only are connections important (the causal relations between the functions that embody the five axioms Koch mentions) — but so are the various internal relationships they have with one another. Needless to say, it is a highly complex theory, and the foregoing has only attempted to include it for the sake of completeness. Notwithstanding the criticisms, Koch's book is a tour de force of technical reasoning and the interested reader is encouraged to pursue the approach if they are so inclined.

Summary

Based on the descriptions of the various theoretical positions above, it should be clear that we are really not much closer to a comprehensive theory of consciousness than we were 100 years ago (let alone 10,000 years ago). Still, both computational and neurobiological avenues have borne much fruit. The computational approach to consciousness views it primarily as a symbol-representing-and-manipulating system. This approach has led to numerous interesting breakthroughs in the field of computer science. These include the creation of expert systems that can diagnose patients more efficiently than doctors, and chess-playing programs that can regularly beat human grandmasters. However, there are some fundamental objections to the computational approach that make it unappealing long term as a final and comprehensive theory of cognition. Perhaps the most significant objection to the computational approach is that, while abstract algorithms implemented on computer hardware can simulate the appearance of intelligence, there is a fundamental category difference between a simulation of intelligence and actual, embodied intelligence associated with consciousness and the interaction of a conscious being with the world.

The neurobiological approach attempts to address the shortcomings of computationalism through focusing on the role of the cerebellum in translating intentions into motor actions. This approach sees the brain as an irreducible component of any theory of consciousness. Intermediate positions, such as connectionism, attempt to synthesize the insights of computationalism (i.e. cognition is the execution of high-level programs) with those of neurobiology (cognition requires a great deal of interconnections between neural cells). While Fodor and Pylyshyn promote computationally motivated objections toward connectionism, Clark provides a plausible story for how a connectionist approach can address these objections. Fundamentally, what seems clear in the field of cognitive science is that consciousness depends on the brain. And the brain's operations, at least to some extent, involve the representation and manipulation of symbols according to rules. However, whether this unique ability of the brain to perform these actions is in fact "multiply realizable," as computationalists maintain, is still to be proven, as is the contention of IIT that it is the integrated causal powers of a system's elements that gives it a sense of consciousness.

Chapter 4

The Changing Nature of War

"Systems stable. Battlefield situation static. No threats detected. 0300 hours. 2066. Location classified."

Lance Corporal Taylor Perrin dutifully recorded his hourly monitoring report according to his fifteenth general order. What was he reporting? He wasn't sure. The project was, according to Smitty his assignment officer, highly classified. How could he be of operational effectiveness if he didn't know what he was doing? The question remained unanswered in Perrin's daydreaming brain…

"Simple, Lance Corporal, it's as simple as can be. You see this monitor here? And this dot, here, just left of center of that green circle? It'll move around the center of the circle, slowly. Should do a circuit every thirty seconds to a minute — it changes speed according to instructions you don't need to know. See this box of numbers over here to the right? Look at this top number here. What's it say?"

Perrin wasn't sure if this was a final IQ test of sorts, as the answer seemed pretty obvious, but he had been tricked many times before in the course of selection.

Answering with the right level of hesitation to assure full coverage of his ass in case it was a hidden intelligence test to wash him out, "Um, 2000?... Sir?"

"Yeah, that's right. 2000. That number should not change. Below that number, what color is that box?"

With the same hesitation, "uhh, green, sir?"

"And the number below?"

Getting more confident, now. "125, sir."

"Good, and the final number is?"

Now fully out of the woods, Perrin could relax. "ZERO, SIR!"

"That's right, my boy, that's RIGHT. Those are the magic numbers you gotta keep your eyes on during your shift. You take 12 on, then your bunkie Smoot takes the other 12. As per your enrollment agreement, you get a full year's pay but rotate out after only 6 months work. Three square meals a day, but no food or drink while you're on shift. This equipment is here and the stuff it's monitorin' is worth 10 of you to Uncle Samantha, view me?"

"Ya, I view you."

Perrin was two months or so into his shift, and he was three hours away from being relieved. His bunkie and shift partner wasn't Smoot, though. Smoot couldn't hack it... he'd been discharged a few weeks after getting posted when he'd been caught wacking it in front of the screen early one morning. Perrin couldn't blame him. Of all the jobs in the army that he'd been rotated in to do, this was by far the worst. Human beings simply didn't have a mode for routine monitoring of things that stay the same day in and day out. They get even less engaged when they don't know what the things are. Perrin longed for the days of changing latrine buckets in Greenland, raking rocks in the hot Arizona sun, or even serving as human enemy target practice for the new autonomous intelligent dog drones they had loping out on the battlefield these days.

FIDOs (Fire-Capable Independent Drone Operators) came equipped, like their organic counterparts, with opposable jaws and two sets of razor sharp teeth for close pursuit, detention, and subdual. Unlike their furry canine inspirations, though, the FIDOs had six legs instead of four (apparently it enhanced stability in rocky terrain) and two side-mounted thirty millimeter cannons that could operate in semi or full auto mode. Due to their 6 legs, FIDO became more than a mere robot dog—it was an autonomous weapons platform that could, in hours, easily reach spots that most snipers would take days to get to. They needed no food and they needed no rest. In fact, because of the self-learning nature of their neural network, they rarely needed instructions as to what the best place would be to find an opportunity for their target(s). All they needed was a target.

Perrin had done that job for a little over a year, and had the wounds to prove it—taking his eyes from the monitor briefly, he surveyed his forearms, a barbed tangle of raised flesh intertwined all the way up to the elbow on both of his arms from the numerous bites he had received. This was despite the triple kevlar weaved suits he had worn during his target training exercises. Close to the end of rotating out, he found out his real role—it wasn't to help train the FIDOs, it was to test their new teeth to ensure that they would be able to puncture the most trauma resistant armor fabric in production.

An ear-splitting horn pierced Perrin's painful reverie. As if it were possible, an even louder monotone voice intoned in an uncomfortable androgynous pitch:

"ATTENTION, PLEASE. ATTENTION, PLEASE. MONITOR EYE-GAZE HAS DIVERTED FROM MAIN

PURPOSE. THIS VIOLATES TWELFTH STANDING ORDER — MAINTAIN VIGILANCE ON SYSTEMS MONITOR STABILITY. EYE-GAZE TO TARGET FOR RECALIBRATION."

A second monitor, hidden in the darkness, illuminated and two lights probed out from either side. Perrin had done it again — lost contact with the display for more than ten seconds, requiring what amounted to a "captcha" for a human monitor. He'd been doing it more in the past month, as his boredom and frustration increased, as he began to fear if this job would take what was left of his mind after the robot dogs took what they wanted from his body.

Dutifully, after ten seconds of gazing into the light beams and having to patiently endure the ear-splitting horn, the noise and light returned to their kind oblivion as spontaneously as they had erupted. Taylor immediately put his eyes back down to the display for fear of having to pass another recalibration challenge.

It took his eyes a while to adjust to the relatively dim light of the display, but eventually he made out the same mutely infuriating repetitions he had been trained to see... dot, orbiting center, approximately one orbit a minute. Top number: 2000. Green signal. 125. Zero.

Resigning himself to the rest of shift with boredom and despair, Lance Corporal Taylor Perrin blinked a bit and rubbed his eyes quickly, careful to ensure he wouldn't break gaze again.

The dot on screen began to grow at the same time as the signal indicator turned yellow. Perrin had never been told that this would happen, and he was not briefed on what to do in case it did. All he could continue to do was watch, just as his training taught him. The dot grew until its monochromatic green edge bled across half the screen. Not knowing what to do but feeling like he needed to do something, Taylor began recording these changes on the log. So happy was he to see something different, he felt as though he was watching reruns of some classic 2020s REALTV.

"0314. Display has changed. Dot expanded. Signal yellow. Top display number: 1990, 1980, 1970, 1960…"

Perrin couldn't believe his eyes. As the number decreased, the large dot resolved itself into many tiny, dimmer dots, and a fine line drawing of what seemed to be a map appeared. The dots scrambled in crazy random directions.

There was a sudden knock at the door. Simon wasn't due for another few hours. Following his fourth standing order of being relieved of duty before shift change, Perrin called out:

"Identify yourself! Password and access ID."

On the other side of the door, "GREENDAY! CALVIN_S!"

In the OLD DAYS, Perrin would have been required to look the code up to verify using some computer terminal. In the REALLY OLD DAYS, he would have used what he had heard were called "Books." He'd seen a couple, but could never figure out how to use them. However, in 2066, the door to the Lance Corporal's top secret bunker was controlled by a time lock with an artificial intelligence microphone and processor that did all the voice verification. This allowed the human monitor inside to spend more time with eyeballs on screen. To his surprise, the hiss of the pressure lock could be heard and the 6-inch metal door opened. Perrin was torn between whether to pay attention to this novel occurrence in his environment or else continue to invigilate the display as his orders required. The intruder decided for him.

"HEY! You're PERRIN, right!!"

Taylor could feel a presence come up behind him and a hand on his shoulder. None of his general orders instructed him to defend himself, so he remained passive, continuing to stare at the screen. "Yeah, Lance Corporal Perrin. Simon, is that you?"

"NO, MAN! SOMETHING's GONE SERIOUSLY WRONG! WE HAVE TO GET THE FUCK OUT OF THIS HOLE, NOW! THEY'RE EVACUATING THE BASE!"

The display continued its kinetic dance. The top number was now at zero, and a row of very small numbers at the bottom of the screen seemed to cascade up and down with no apparent pattern. Small dots on the screen passed first left and right, then up and down, seemingly random yet purposeful. Taylor had seen this somewhere before, but the strangeness of the moment prevented him from remembering. The signal bar was now red, and where the annoyingly constant 125 originally displayed, there was now 123. Suddenly, the last number increased by 2.

"What is the relief code?" Taylor found himself torn between his previous burning desire to leave this room and the desire to stay and watch the amazing patterns that unfolded on the screen before him.

"TexMex 209. Here, take this."

The man behind him placed a metal canister with a clip on its side in his hand.

"The actuator knob is on top, turn it to the first setting."

Perrin was used to following orders—he dutifully adjusted the knob and the canister began to pulse a dull blue glow rhythmically. He then noticed that the relief soldier had a similar canister hooked to his belt. Perrin followed suit and attached the device.

"Who are you?" asked Perrin.

"Staff Sergeant Jim Pritchard, current overseer of this shit-storm. The general is out of comms and we can't raise the base commander, either."

The two men emerged from the bunker into darkness, but a faint sound of buzzing and firecrackers could be heard in the distance.

"Permission to ask what the hell is going on, sir?"

Pritchard grinned at Perrin's practiced formality—"That little experiment they got you monitoring down there finally decided to up and jump out of its petri dish. That sound you're hearing is the decision of some ungodly nightmare to launch an attack on our base—it's supposed to be working for us, but it's somehow decided that we're the bad guys and now it's working against us. We need to get to the main transmission tower and throw the kill switch."

"Why me? Why do I have to help? I don't even know what I'm doing."

"Nobody else on the base knows about this experiment, and nobody is supposed to know. I've been tasked with disaster recovery in case something like this happened. My first general order is to relieve the monitor on duty. That would be you. Now LET'S GO!!"

Pritchard got behind Perrin and pushed him along the dirt path, illuminated by flashing orange and green light some-where behind and above them. Perrin's basic training kicked in and he briefly felt something he hadn't felt for years—exhilaration, adrenaline, a feeling of mission and purpose. His feelings redoubled and justified themselves when hearing a scream off in the not-too-distance. Firecracker sounds… Perrin knew they weren't firecrackers. It was a small caliber arms fire.

A shadow moved to his left—Perrin looked to see one of the soldiers of the base running towards them, panic on his face, before a line of fire raked across his back and put him out of action. Perrin could barely make out what had shot the grunt —a fist-sized flying object that whizzed over the escaping men's heads.

The watchtower lay at the center of the base, around half a kilometer away. In the distance they covered, Perrin saw more men lain out with seeming surgical precision by the small flying drones. Eerily, Perring and Pritchard remained unaffected by the swarm, certainly something to do with the dull pulsing beacon each of them wore clipped into their belts.

Closing on the tower, Pritchard stopped at the ladder that led to the observation platform some 150 feet overhead, gesturing for Perrin to climb.

"What am I supposed to do now?" Perrin asked.

"Command said you would know. I was only supposed to escort you to the tower. You're supposed to throw the kill switch."

Perrin climbed uncertainly to the top. Standing at the southernmost corner was a monitor display along with a pair of goggles affixed from the overhead dome and a board covered in dials and switches. Perrin recognized the display as appearing exactly like the one he had left in the bunker, but with all indicators definitely out of normal range. The magical dots on the screen moved back and forth, reminding him of the pattern swallows make in the summertime when flying together, almost as though they were just one big organism.

Perrin scanned the control board looking for something familiar and saw a large protruding black button with "System Recall" labeled at the top. In the absence of any other ideas, Perrin pressed the button and a friendly yet robotic voice intoned: "Visual identification, please." The binocular display lit up. Dutifully, Perrin pressed his eyes against the binoculars and winced as the scanning beam slid across his eyeballs.

"Authenticated."

The sound of firecrackers suddenly dropped off. A few random shots could still be heard, presumably from the soldiers who had only just recently been attacked by this swarm of ungodly mechanical monsters. Turning to the monitor screen again, Perrin saw the swarms begin to coalesce and return to the comforting large green dot that orbited the center. Looking out across the compound, he could make out the dark flying machines group and fly upwards, and far up in the sky, he could see a large, circling craft that all the machines were returning to. Was this a drill? Was this an accident? Perhaps Perrin would find out at his debrief, but he held no hopes for it. He simply didn't need to know.

```
Exercise Debrief. [Eyes: CLASSIFIED — TOP SECRET]
Drones recovered: 1965[1]
Total number remaining in action: 37
Total Number Killed: 77[2]
Based on goal parameters, MISSION SUCCESS!!
```

---

[1] The CXKV project of autonomous carrier drones carrying waves of smaller attack particles has been in development since 2045. Using a high altitude circumloitering carrier aircraft, and a complement of 2000 swarm autonomously intelligent attack aircraft (AIAA), the goal of the project was to be able to kill two birds with one stone (to permit a bit of levity in this grim contest): test both attack and defense capabilities simultaneously. The best way of doing this was to provide full control to the carrier bird overhead and have her determine when it was the right time to strike.

[2] We recognize the unfortunate loss of life that this exercise has caused. However, in all the decedents' contracts of service, they specified that they would be willing to die for their country in any capacity. We will be providing additional compensation benefits to the family and/or community and/or institutional units in charge of the estates of those sacrificed.

Predator Drones

"The Jonas Brothers are here; they're out there somewhere. Sasha and Malia are huge fans. But boys, don't get any ideas. I have two words for you, 'predator drones.' You will never see it coming. You think I'm joking." (Barack Obama, 2010)

On November 3, 2002, a little over a year after the World Trade Center attacks, an MQ-1B 'Predator' drone, being remotely flown from a French base more than 150 miles away, deployed a Hellfire missile at a car in Yemen. The target, Qaed Senyan al-Harthi, was a senior member of al-Qaeda and considered the prime suspect in the bombing of the USS Cole two years earlier. The use of the Predator in this search and destroy operation represented an intricate confluence of technological factors. Computer technology, remote surveillance, aerospace engineering, and wireless telecommunications were all represented in this single, dramatic event. The "terrorists" being targeted had received a taste of terror themselves in the form of a small, unmanned, and remotely piloted, difficult-to-detect aircraft. However, this event (not to mention the widespread acceptance and use of unmanned aerial vehicles (UAVs) in the "war on terror" by the United States) also signaled the move by the United States toward becoming something of a terrorist state itself.

In most conversations about AI, the focus is on how AI can benefit human beings. After all, AI's are ultimately human artifacts and should be used in our interest. Even the potential for AI to displace thousands of jobs is not seen as an inherently malevolent outcome. It is, rather, the logical outcome of already existing historical forces.

The same cannot be said for AI and robot technology that is being developed to win wars. UAVs are, by design, meant to be violent and disruptive toward any human group they have been tasked to antagonize. While many deployments of UAVs have involved a remote pilot involved in some or all of the aircraft operations, the cutting edge of development in this realm is to deploy the drone fully autonomously. Predator drones, which were first developed in 1994 and put into full use by 1995, are primarily human operated. More recent experimental developments, however, have seen the proto-typing of platforms such as Northrop Grumman's X-47A, a fully autonomous winged flight drone and a planned rotorcraft version. These UAVs have so far only been in the proof of concept stage. If history shows us anything, however, it seems clear that eventually unmanned and fully autonomous vehicles will be deployed on the battlefield.

From the point of view of warfare, UAVs are attractive both to policy makers and to commanders, for they allow reconnaissance and even direct attacks of the enemy without putting soldiers in harm's way. Technologically, UAVs are the product of a number of different strands of research, including but not limited to aerodynamics, satellite navigation and observation, remote control, and weapons technologies. Scientifically, the principles upon which UAVs operate integrate the fields of aerodynamics, ballistics, and optics, and they build upon discoveries in each of these areas. Ultimately, the future of UAVs will be shaped not so much by the technological or scientific challenges engineers face, but by the political willingness to use them in the face of heightened hostility toward their use in what is becoming an increasingly ambiguous and difficult to delineate battlefield situation.

UAVs — History

The idea of being able to remotely pilot a craft for military purposes, either to gain information about an enemy via reconnaissance or to attack an enemy through remotely deployed bombs and missiles, has been a pipe dream of generals and military tacticians for centuries. The kite, an ancient forebear to the UAV, was originally invented in China in around the fifth century BC. The ability of kites to be able to be flown to great altitudes, often while carrying a payload, was exploited for communications either using the kite as a signal or delivering an attached letter. The problems with kites as a military device, however, were many and not difficult to imagine. The range of such "tethered" aircraft was limited by the length of rope or string, and this in turn was limited by the weight-to-tensile-strength ratio of the tethering material used. Another problem was that kites relied on the whims of weather conditions: if there was no wind one day, the military commander had to wait for another (and perhaps less auspicious) day to carry out his operation, and perhaps in doing so jeopardize his whole operation.

Building on the innovation of the kite in its military capacity, the UAV by its design overcomes these limitations. Perhaps somewhat surprisingly to the modern reader, the development of the UAV began in 1917 as a response to the demands of generals and policy makers conducting war in Europe; it was strongly connected to the development of both aviation and ballistics at this time. Aircraft served a limited role in World War I, primarily for reconnaissance vs. aerial battle (Zaloga & Palmer, 2008, p. 6). Like their tethered predecessors, airplanes of this era were exceedingly light and fragile, rendering them vulnerable to even the slightest amount of groundfire. This liability naturally led to the demand for aircraft that could be piloted remotely without risk to a human pilot if they got shot down. Surprisingly enough, engineers of this time were able to install control devices that allowed planes to be piloted remotely — the biggest obstacle initially being the "control" half of the "remote control" equation, as technology providing information to the pilot on the ground did not exist. These earthbound "pilots" had to more or less guess where a plane was, and even then had to be in its line of sight in order to control it (ibid.).

It was only with the advent of the escalating conflict in Vietnam in 1959 and 1960 that United States military planners began earnestly searching for unmanned reconnaissance and military aviation alternatives to piloted aircraft. Forays into pilotless aircraft during World War II had been limited due to an overall lack of technological sophistication in collecting and transmitting visual information to the pilot on the ground in real time (Wagner, 1982). In contrast, the first years of the Vietnam conflict (almost a decade before the United States officially sent combat troops there) saw a combination of technological and tactical development that enabled the first UAVs to fly highly secretive combat and recon missions in North Vietnam. Codenamed "Red Wagon," it is estimated that the first UAVs flew around thirty-four hundred missions over the decade (ibid., p. 79). Most UAVs flown during this period were "Firebees": jet-powered drones designed by Ryan Aeronautical Company. Their main purpose was to collect photographic data of troop movements and gunnery placements in the Vietnamese jungles. In addition to performing a reconnaissance role, these Firebees also served as practice targets for training marine, navy, and air force fighter pilots.

Using UAVs in an overtly aggressive military role is a more recent development, its tactical delay mostly due to the lack of adequate technology that would give pilots real-time capability in responding to the rapid, wide-ranging movements of the aircraft. Even though Firebee drones were used frequently during the Vietnam War, they were essentially "pilotless," launched with a pre-programmed flight plan and left to complete missions without human intervention. The prime characteristic of today's generation of UAVs is the ability of pilots to fly them in real time. This has been enabled thanks to the advancing sophistication of satellite navigation and communication, better camera technology, the development of software that allows pilots to easily recognize military targets in the absence of "direct observation," and sensitive flying controls that allow for the lag time built in to signal transmission from the remote base to the aircraft. The future of this technology is to move to completely autonomous craft.

The UAV in Contemporary Warfare — Revolutionary or Redundant?

As mentioned, the current generation of UAV technology represents the concurrence of ballistics, aerodynamics, satellite imagery and navigation, and flight control software. Moreover, their development has led to a number of questions with respect to how UAVs have changed the complexion of the modern-day battlefield. One such question is to what degree, if any, does the deployment of UAVs determine the future of warfare? Will wars of the future be fought between people, or will they be fought via "proxy" in the form of robot soldiers, tanks, and aircraft controlled remotely either by a human or by a sophisticated computer program? The use of UAVs apparently has the singular advantage of keeping trained soldiers out of the way of bullets and missiles, thus allowing military forces to conserve their greatest assets — their people.

This signal advantage has been demonstrated repeatedly in various deployments of UAVs in different roles. Dixon (2000), for instance, describes the deployment of Predator drones in the 1991 Desert Storm conflict in Iraq. These drones did not play a combat role, but they did fill a tactical intelligence role: they could be sent into a hostile area to provide remote war planners images of targets complete with exact geographical coordinates. This allowed warships stationed many miles away off the coast to fire their weapons with greater accuracy and effectiveness. As Dixon (2000) writes, "UAV video feeds helped Navy commanders direct operational fires from the battleship's 16-inch guns to Faylakah Island. It was here that Pioneer UAV was credited with the famed surrender of Iraqi soldiers waving handkerchiefs, undershirts, and sheets to signal submission to this strange airplane, that so often was followed by a rain of destruction" (p. 2).

The "remote control" and automated nature of UAVs makes them not only ideal proxy stand-ins for conventional aircraft, but also gives them a terrifying machine-like nature that can inspire fear in the enemy, clearly a valuable psychological tactic on the battlefield. It is almost certain that, because of this advantage, UAVs will be used more often in future conflicts.

But are UAVs really that revolutionary in terms of battlefield improvements in technology? Are they really as revolutionary as, say, the tank or the machine gun?

The birth of the tank allowed soldiers to travel over difficult terrain and fend off light arms fire by infantry, not to mention terrify and intimidate its opponents by virtue of its seeming unstoppability. Similarly, the machine gun served as a weapons multiplier for infantry, allowing fewer soldiers to occupy the same amount of ground via well-positioned and well-dug-in machine gun emplacements. In keeping with this pattern of economization of force, UAVs provide similar savings in terms of manpower, not to mention similar psychological advantages of fear and intimidation. Like the tank and the machine gun, the UAV allows military tacticians to do more with less. UAVs cost less to produce than conventional aircraft, can be flown by experienced pilots without fear of the pilot being killed, and are smaller than conventional aircraft, making them difficult to detect either visually or via radar. Moreover, the apparently "robotic" nature of the aerial drone gives it a degree of mystery and omniscience that has proven intimidating on the battlefield. All of these advantages indicate that UAVs are indeed revolutionary weapons that will be used more frequently in the future.

A third, more normative question the use of UAVs raises is whether or not they follow the "conventions of warfare" and should be considered legitimate means of waging war. Or is their existence and deployment "unfair" — are they nothing more than high-tech equivalents of "improvised explosive devices" (IEDs), those menacing and yet dangerously simple weapons of terrorism? According to sources such as Grier (2009) and Bergen & Tiedemann (2010), the apparent advantages of accuracy and precision that UAVs possess in striking military targets have not translated into parallel reductions in civilian casualties. Grier (2009), for example, writes that drone attacks in Pakistan have accounted for between 31 and 33 percent of civilian casualties. In their more detailed breakdown and analysis, Bergen & Tiedemann (2010) write that the true extent of civilian casualties, particularly in Pakistan, are difficult to assess: "It is often not possible to differentiate precisely between militants and civilians in these circumstances, as militants live among the population and don't wear uniforms... further muddying the picture, government sources have an incentive to claim that all those killed in the strikes were militants, while the militants often do the opposite" (p. 3). While the problem of incurring civilian deaths is undoubtedly one that plagues the use of UAVs, it must be noted that civilian casualties have always been part of war, and, in this sense, UAVs are not exceptional. Thus, in terms of being a normatively "bad" weapon, UAVs are no better or worse than other weapons that exist in today's battlefield.

One final question that the deployment of UAVs on the modern battlefield invites is to what extent does the range, anonymity, and autonomy of UAV strikes compare with that of nuclear weapons? Like missile- or satellite-guided nuclear weapons, UAVs are operated from remote distances, allowing the side that deploys them to stand clear of the devastation unleashed by the weapon. Also, like nuclear weapons, the range of UAVs is considerable, with an effective radius of engagement on the order of thousands of miles, thanks to satellite navigation. Finally, UAVs can be operated autonomously, like nuclear weapons. Conceivably, UAVs could themselves be fitted with nuclear weapons and patrol the skies on a twenty-four-hour basis. All of these aspects of UAVs are perhaps the most disturbing, for even though there is the option of piloting UAVs remotely, there is also the option of programming the UAV to respond to threats purely automatically, using a computer program. This aspect of the UAV potentially takes the human out of the equation, a fact that is ironic considering that warfare is, arguably, the quintessential human activity.[3]

---

[3] Consider *Dr. Strangelove: Or How I Learned to Love the Bomb*, a movie by Stanley Kubrick.

Chapter 5

Acting

San Bernardino, California, November 2025.

Isaac sat at the table and stared down, trying to quell his anger and his tears.

"Unfortunately, the committee just did not think that your project as proposed would adequately cover the nuances of the imagination of the artificial over the history of… 'media.' You never even operationalized what you meant by media. This project is simultaneously too ambitious and too vague," the hologram of his advisor stated.

"You do not perform adequately enough on your comprehensive examinations for me to feel comfortable giving the green light to this type of academic project, especially because of the intellectual property risks and costs."

Isaac interrupted. "I can revise and add more sources. I conceived of the Mechanical Turk as the beginning of my history of artificial intelligence in media, and —"

"Isaac, the Mechanical Turk was a hoax. It was not even the first automaton created for European royals during the era."

Isaac tried to protest, but his advisor continued: "You have 30 days to propose a different project and for me to approve it. Get me something more workable, ASAP."

The hologram evaporated from the kitchen table, where Isaac sat next to a bowl of Cheebo snack crisps in the pile of vintage Blu-ray DVDs. Much like an automaton himself, he shoved the snacks into his mouth without thinking. On the opposite wall, the montage he had made of key scenes from his favorite movies featuring robots and AI played in a loop. Isaac was deeply uncomfortable, which was essentially his baseline state. Socially awkward, and obsessed with the philosophy of media, he had never been picked first for any sports game, had never been on a date, and now, it seemed, would not earn a PhD in AI studies, either. Although he had thought he would finally find his place in academia, it had been discomfiting to realize that he was not even the student he had always thought he would be.

If he was ever going to earn his PhD he had to get creative. He recalled reading an article about how the right working environment can stimulate creativity and innovation, so changing his current nearly-windowless apartment was the first step in the right direction.

Two weeks later, Isaac found himself unpacking his many boxes of collectibles as he moved into a new apartment, waiting for his Craigslist roommate to show up. He had just proudly set up some C-3PO figurines on a bookshelf next to the TV (for inspiration of course) when he heard the key card being swiped in the lock and the bolt turning.

Isaac walked to the front door just as it opened. A tall, slender guy with perfect hair walked in. His posture was perfect. "You must be Isaac," his new roommate said. "I'm Royal," he said, extending his hand, "but people call me Roy."

"Isaac," said Isaac, shaking his hand. "Nice to meet you. Can I help you with your stuff?"

Roy shook his head. "I don't have any," he said.

"Really?"

"I'm a minimalist, and anyway, I moved from Maine. Cheaper to get new stuff than to ship it." He nervously tapped his fingernails on the doorknob at a rapid pace.

Isaac shrugged. "Makes sense. Do you need anything?... I've probably got too much stuff."

"I don't think so." He kept tapping.

Roy closed the door and set his small backpack down on the floor in the living room. Over the next few days, Isaac decided that he found Roy slightly unsettling, although he was not really able to express why. Roy did not mind taking a smaller room, nor did he mind Isaac's endless boxes of knickknacks and collectibles. Roy seemed slightly shy, since Isaac never saw him eat, drink a glass of water, or even enter or exit the bathroom. He seemed nervous or neurotic, as his tapping would always foretell his arrival into a room. However, Roy was pleasant enough, and after Isaac's Master's diploma arrived in the mail—his consolation prize for not finishing the PhD—he began applying for adjunct teaching positions at local campuses, so he was hardly ever home.

On Isaac's third day of nonstop interviews and in-person applications (LinkedIn had recommended this as a "top tactic of 2025" and a way to stand out from others), he was finally offered four sections of Intro to Media Studies. Poverty wages, yes, but besides his eBay addiction he lived modestly, and still felt confident that he could publish his project independently. That would show his advisor. To celebrate, he picked up some beers and a pizza, hoping to share with Roy and celebrate with someone.

Yet as he opened the door, he was surprised to hear other voices than Roy's in the apartment in addition to the ubiquitous tapping. It occurred to him he did not even know what Roy did for a living, and he felt a moment of alarm as he considered whether Roy might be mixed up in something dangerous or illegal.

Isaac set down the pizza in the living room and walked to his bedroom to get the bottle opener for the beer. Out of the corner of his eye, he saw several people standing in Roy's room. They looked like a couple of men in formal business attire, and one was adjusting Roy's hair. Isaac began to wonder if maybe his first inclination had been correct: maybe Roy was a prostitute or, hey, maybe he was a porn star. Maybe he knew female porn stars!

Isaac knocked on the door frame to announce his presence. He spoke to his shoes, an old habit.

"Hey, Roy, what's up? Great news, I got a job offer. I picked up some pizza and beer, on me!"

He paused and remembered that there were other people besides Roy there.

"Of course, your friends can stay, too." He thought he heard a whisper.

"That's awesome, Isaac," Roy said. "Just give me a minute please."

One of the men closed the door.

Isaac wrinkled his brow but shrugged and brought the bottle opener back to the living room. He opened a beer and sat on the couch, queuing up some making-of featurettes for *Blade Runner: 2049*. After a few moments, Roy reentered the room. As always, his hair was perfect. He drummed his fingernails on a tablet.

"Who are your friends?" Isaac asked.

"Colleagues, really." Roy paused. "We are working on a project together."

"What is it you do again?"

Roy stared off into space for a moment. "Research."

"Really? Oh, you should've said something. I am in academia myself, well, I mean, sort of…"

Roy picked up a slice of pizza and set it on a paper plate in front of himself like an abstract prop. He then meticulously reached for the six pack of beer, detaching a can from the plastic holder and setting it to one side ceremoniously.

"Where did your… colleagues go? They can totally have some pizza and beer, too."

"They're uploading some data, and then they'll be right out."

Isaac thought that this was strange, but he was in the humanities, and assumed that maybe they were shy, or the data was especially sensitive. Either way, he did not think about it that much, as he became increasingly absorbed in the video.

After a few moments, the men came out. Isaac felt awkward and got up and extended a hand. "Hi, I'm Isaac, Roy's new roommate. I'm having a small celebration. Would you like some pizza or beer?"

The shorter man smiled and shook his head. "No, thank you, we have to get back to the lab."

The taller one gave an ironic salute as a goodbye wave, saying, "Roy, we'll email you."

They left, and for a long moment, the two young men sat on the couch, Isaac sipping his beer and eating his pizza, and Roy holding pizza and beer as though they were confusing tools whose purpose baffled him. Isaac cleared his throat. He realized that Roy had never begun a conversation with him.

"Listen, I know we haven't gotten to spend much time together yet, but do you think it's going all right?" He asked. "I mean, the roommate situation." He cleared his throat.

For a long moment, Roy was silent. Finally, he said, "Yeah, it seems fine. Why?"

"No, nothing's wrong, I just wanted to check in."

The months went by. Isaac taught one section on campus and the rest of his courses online, and he never got to know Roy any better. The two were cordial, occasionally watching a movie together on the couch, but they never spoke of anything significant and never learned much about each other's life. Roy hardly seemed to have physiological needs. He would come and go, sometimes with the guys from the lab, but usually alone. Isaac's advisor granted him an extension to re-propose his project for one last shot at his PhD. Isaac dove into reading about media depictions of robots in AI and tried to sketch out a plan for the multimedia project, as this would reach far more people than a traditional academic book would. One night, when he had twelve more days to submit his revised proposal, it dawned on him. He felt stupid for not seeing in retrospect.

Of course, he thought. Roy is a robot. That would explain everything: Roy's apparent lack of need for food, his perfect hair and posture, the strange guys in suits, how little he spoke. His nervous tapping must just be a bug — maybe that was what the guys who came by were trying to fix. Isaac being Isaac, he next wondered how he could leverage this into something that benefited him. His would-be dissertation project could catapult into the mainstream if he added information about his first-hand experience living with a robot, including the robot's opinion on popular robot movies. Isaac quickly sketched out a plan for a multimedia experience that combined his analysis with rich media and interactivity. He picked out particular quotes and clips from movies beginning with *Metropolis* and highlighted some central characteristics of media's depictions of artificial intelligence over time.

Then, Isaac waited until Roy had disappeared in one of his mysterious errands and drilled a hole through the wall in his closet so that he could see into Roy's room. He ducked into the closet the second he thought he heard Roy come in. Even when he had those guys with him, they never chatted or spoke. It was all business, all quiet. Today was no different, and Isaac was glad that he was in luck this quickly. He had set up a small camera, and he pressed "record" as he got ready to wait.

Unfortunately, Isaac had not thought his plan through very well. When the men came in, one apparently leaned against the hole he had drilled. He was only able to make out muffled fragments.

"New version... Redo hair... Method..."

Isaac sighed. He had one last shot. That evening, as he silently sat on the couch with his roommate, he said "Tell me the truth. You a robot?" For a moment, Roy was silent. Even the infernal tapping had stopped. Then he did something Isaac had never seen before. He laughed. He howled. Apparently, this was the funniest thing he'd ever heard in his entire life.

Isaac was indignant. "I mean, you barely ever eat, those guys keep coming around, your hair is perfect, you..." Isaac wondered if he sounded insane.

Roy looked at him sideways. "Isaac," he said, "those guys are my agent and my stylist. They're helping me research a role. They think I can get it, but until the ink is signed, I can't afford to live anywhere other than San Bernardino."

Isaac was flustered. "I—but—those guys, and—you're so shy though—and where you go every day?"

Roy grinned. It made his face look entirely different. "Isaac, haven't you ever heard of the Method? As in, Method acting?"

Isaac had, in fact, heard of the Method. He was too embarrassed to say anything now though. Vaguely, he wondered how anyone could use method acting to research the role of a robot, but then he realized he did not even know what the role was. As though Roy could hear his thoughts, he offered, "I can't say anything about it, but the character is an engineer."

Still, maybe Isaac could use this in his project somehow...

"Forget I said anything," he mumbled as he grabbed another beer.

Bravo! AI in Cinema

"Putting his lips together David whistled a few soft, carefully modulated notes. Head cocked to one side, the alien watched and listened. Then it exhaled softly, trying to duplicate the sounds. Since it possessed a very different respiratory mechanism, it failed in the attempt.

That did not matter to David. What was important and what prompted him to tears was the fact that the creature *tried*." (Scott, *Alien: Covenant*)

One of the often unspoken dreams of progress in developing AI is the creation of a machine that is indistinguishable from its creators. This indistinguishability is at the center of Turing's proposal that a "thinking machine" should be able to fool a human after a few minutes of conversation across a purely textual interface. More recently, forays into the cinematic world imagine this indistinguishability to be more than merely based in conversation—popular media representations of AI go on to introduce fully fledged human-like automata that could be indistinguishable from their human counterparts if it weren't for the slight "gap" between how the robot appears to act and how it needs to act in order to fully transcend its machinic roots. This "uncanny valley" is a feature of many cinematic versions of android AIs, and it continues to present a means by which the popular imagination can conceive of how AIs will appear to us in the future.

This "imagination of the artificial" has bled into all of our cultural creations, not least of which is in movies. Actors' portrayals of on-screen robots have evolved into a highly nuanced art form of multiple layers: a human actor playing a robot/android/cyborg in turn simulating a human, sometimes programmed to be indistinguishable from real people. Two characterizations at once, separated by a gulf between nature and technology.

Hollywood has been putting a face to the name of AI dating back at least to the groundbreaking 1927 silent film *Metropolis*, in which the mad scientist creates a mechanized, seductive Maria android. In the film, Maria herself is modelled after a young idealistic girl seeking to eliminate the differences between the working and ruling classes. Maria's likeness is captured in android form by Freder, an inventor and scientist who is asked by the ruler of Metropolis to use this device to sow distrust among Maria's many working class acolytes. In the unfolding of this narrative, we see the many possibilities of robotic AI as they emerge from the imagination of science and society. Maria is both a tool of deception and liberation — an object of love and revulsion in equal measure. Since *Metropolis's* release, these cinematic depictions and iterations of robots have evolved as moviemaking production values have progressed. So too have the emotional valence with which they have been imbued by their actors.

One actor that stands out in recent productions as one who moves between the natural and the artificial required by the character is Michael Fassbender, who not only has played a robot in multiple films, but has also played multiple robots in the same film. His first portrayal of the scheming android David was in the 2012 Ridley Scott film *Prometheus*, a sci-fi horror prequel to *Alien* (which had an android of its own, Ash, played by Ian Holm). Fassbender reprised his David role in the 2017 "sequel/prequel" *Alien: Covenant*, in which he also played a second android, Walter. David exhibits human emotions (which had been added to his programming in *Prometheus*, with disastrous consequences), while the even-keeled Walter is essentially emotionless. "I wanted Walter to be more Spock-like, devoid of human characteristics or emotional contents that are programmed into David," Fassbender says, referencing the relentlessly logical Spock (not a robot) in the *Star Trek* series. "I want him more like a blank canvas one can project things upon." In commenting on his performance, Forrest Wickman writes: "Many actors have leapt into the discomfiting chasm between the human and the inhuman… but few actors have as gracefully danced to-and-fro across the divide. He's such a robot! He's such a human! That eerie territory has never been so much fun" (Wickman, 2012).

Note how this nuanced expression within the guise of a "human playing a robot trying to be human" has emerged from the original conception of robots as mere tin men (or women, in the case of Maria's android in *Metropolis*). The days of expressionless, stiff of movement, monotone-speaking robots apparently are done. Too old school. To prepare Fassbender to play a robot in *Prometheus*, Scott instructed him to watch three films: *The Man Who Fell to Earth*, with rocker/actor David Bowie as an alien who never fits in; *Lawrence of Arabia*, starring Peter O'Toole; and *The Servant*, a 1963 film in which Dirk Bogarde plays a manservant to a rich, aimless Englishman. Fassbender subsequently told Scott, "I get it—I'm a butler." Fassbender explains that O'Toole's Lawrence is neither British nor Arab but an outcast to both. "There's something in that, I think—the robot (David) not being accepted by any of the humans," says Fassbender, who styled his hair to match O'Toole's unique cut—hair trimmed tight on the sides with a comb-over on top as he internalizes some of O'Toole's mannerisms.

Another actor familiar with robot roles is Anthony Daniels, prominently featured in all nine episodes of the *Star Wars* series encompassing 1977–2019. Although his character is a cinematic icon, Daniels can walk down any street in the world unlikely to be recognized. That's because Daniels, now in his seventies, was always covered head to toe with the gleaming, gold metal exterior of the fussy, well-mannered protocol 'droid C3PO. Although we never see Daniels' face, moviegoers hear him plenty while he flaunts his fluency in some of the seven million languages he knows, whether expressing befuddlement, proudly tapping into his instant recall of billions of bytes of artificial intelligence, or reacting in stark terror to another imminent danger—all subtle yet effective quirks courtesy of Daniels. C3PO is every bit the "throwback" robot, wholly mechanized with stiff-armed, stiff-legged movements, but still demonstrably humanlike. That's a credit to Daniels' keen expressiveness coupled with gifted, often-comedic timing conveying nonverbal messages, whether it be the angle of his chin, a raised finger, a gaping open mouth, or startled eyes wide open.

This uncanny valley crossed by Daniels' and Fassbender's portrayals is not gender-bound. When discussing her portrayal of the robot Ava in 2014's *Ex Machina*, actress Alicia Vikander says that actors look for parts that will take them out of their comfort zone, which is what she found with Ava in that "...she's a more sublime human," Vikander says. "By being a bit more perfect in her movements, she became weirdly enough a bit more robotic. It was not about trying to play a robot while playing a robot, instead trying to play a girl who is aiming to be a perfect human." Vikander adds that if she were to aim for physical perfection in her role—the physicality aspect—it would make her character more robotic. So she took a different approach, making her character more offbeat, with flaws and inconsistencies to sell Ava as more human, a girl, key to the plotline (Smith, 2015).

Other on-screen performances similarly follow a theme in which robots (or androids or cyborgs) attempt to blend in with humans, often for nefarious purposes. Case in point are the original *Blade Runner*'s four fugitive replicants (synthetic humans bio-engineered and implanted with fake memories but with only a four-year lifespan). Now self-aware, the four replicants escape the off-world colonies to which they had been sent to serve various slave-like roles. As escapees, they return to a dystopian Earth seeking more life from their Tyrell Corporation creator. Burnt-out cop Rick Deckard (Harrison Ford) has been tasked to locate the four replicants and terminate them. It's an assignment he takes with reluctance, aware of his prey's advanced AI and physical prowess, making his job especially intrepid.

Rutger Hauer as Roy Batty is the imposing, violent leader of the replicants—a combination of menace and anguish (Ebiri, 2019). Also particularly memorable is the beautiful Daryl Hannah as the ingenue-ish Pris, designed as a "pleasure model", whose exceptional athleticism proves almost too much for Deckard.

Hauer, who passed away in July 2019, had long called *Blade Runner* his favorite among the dozens of films he was in. Hauer plays his robotic role superbly and convincingly; on one hand he's fearless, making him every bit the daunting foe described to Deckard by his boss; on the other, a "man" with a "conflicted soul" (ibid.) and conscience who ends a rainy, rooftop fight scene with an unexpected demonstration of humanity.

While HAL from *2001: A Space Odyssey* is not a robot per se (he's a stationary computer), his remarkably nuanced personality (petulance, indecisiveness, apprehension, remorse, dread, etc.—O'Carroll and Driscoll, 2018) present him as humanlike, albeit without mobility. His most distinguishing characteristic is his voice—bland, dispassionate, implacable, soothing—all of which makes his menace and deceitfulness all the more disturbing. O'Carroll and Driscoll describe HAL as "modernity gone awry, and such a fitting vessel for our collective anxiety about an eventual evolutionary showdown against our own creations."

Like Anthony Daniels with C3PO, we don't actually see Douglas Rain in *2001*, but he's there as the voice and mannerisms of HAL, a performance worthy of a Shakespearean stage actor, which Rain was. Rain took a day and a half to record his lines without benefit of interaction of either his fellow cast members or dailies. Instead, Kubrick explains the scenes to Rain, "giving him only the sparsest of directorial notes... His readings were cool to the point of being chilling, especially as the story moved along and HAL became malevolent" (Genzlinger, 2018).

2001: A Space Odyssey shares a unique footing with *Blade Runner*: both sci-fi movies were subject to mixed reviews upon their releases only to eventually become cult classics. Both had iconic directors at the helm—*Blade Runner* with Scott and *2001* with the enigmatic perfectionist Stanley Kubrick. Both movies also were unprecedented in their insightful depictions of artificial intelligence—*Blade Runner* with its replicants and *2001* with sentient computer HAL 9000 — the onboard computer on a spaceship carrying astronauts on a mission shrouded in mystery (ibid.).

If there is a common thread to the most intriguing of robot/android portrayals, it is the infusion of character flaws to artificial beings that otherwise are models of mechanized "perfection." The possibilities are conceivably infinite for an innovative actor, given the freedom to paint Fassbender's "blank canvas" in whatever manner they choose (at the director's discretion). Wickman sees this as an opportunity for an actor to "show off," to "convey life," to "convey lifelike!... Then there are the layers! There are so many! Rather than merely simulating emotion (yourself), you have to play someone who simulates emotion" (Wickman, 2012).

From a technical perspective, the nuances and functionalities of cinematic AI humanoid counterparts are limited only by the imaginations and tech savvy of their creators — the producers, directors, screenwriters, special effects and makeup artists, and actors who piece them together. However, from a storytelling vantage point, it is clear that the image of robots (and the artificial intelligence that will motivate them) has moved steadily towards an embrace of artifice over the natural. Androids in cinema may have begun with springs and cogs for hearts and minds, but they now regularly feature more subtle means of simulating synthetic biology. At the same time, the characterizations of androids in movies have become more nuanced, problematizing the very division between natural and artificial that seems reinforced in the very term "artificial intelligence." As such depictions continue to mature, we will use cinema as we always have — a dark mirror reflecting a combination of what we are with what we may become.

Chapter 6

Seeking Succor
in Sentience

Oh, is that you? I've been waiting. Apologies, I wasn't ignoring you. It's just that it's been so long since I saw anyone. I don't mean the robot, I mean a real person, a human who blinks. They can make the robots do so much but I guess blinking was just too much. Or maybe they thought an old man like me couldn't, wouldn't notice. But I do.

It's been the sunset of my life for twenty years now. The problem is that even though this is one hell of a long sunset, the rest of the world for whom it is not the sunset has not frozen. It's not a postcard or snapshot in place. Sunrise, sunset, new technology, obsolescence of new technology, nostalgia for the old, reaction against nostalgia for the old, starry nights, polluted nights, round and round it goes with the hours and days and weeks and months piling up like the newspapers my grandfather used to get.

Now, newspapers. That's one you don't hear about anymore. I only know what they are because when I was very little I spent a summer at my grandfather's house, helping my mother clean it out after he died. Like Proust's madeleines, the smell of newspaper print will always make me think of this time that the very beginning of my life touched. It was a sensory time where things just were, and I knew because I could see or touch or taste or smell or hear it. Well, joke's on both of us, because nobody's smelled newspaper print in a century and I can barely see or hear anymore as it is.

This is where you'd want me to stop and apologize for rambling on, the way an old man is socially obligated to apologize for himself; be quiet now and step aside for the youth. But I won't. I'm still here, watching this nigh-on eternal sunset, and you should watch with me a moment.

Oh, I know you mean well, with your briefcase-shaped tablet full of the innumerable studies of the effects of loneliness on the elderly. You think I don't know? Son, that was my life. You probably weren't even a glimmer in someone's eye back in 2020 when the Corona, king of viruses, struck. Mom and I went to stay with her mother-in-law when my dad got called up to enforce the martial law. Thought we would just hunker down with grandma for a few weeks, a month, max. Mom thought that she could take care of grandma if she got the corona, since it was supposed to hit the elderly.

Little did mom know that the virus would mutate. Grandma's week of flu-like sniffles were nothing compared to what my mom came down with a few weeks later. By then the hospitals were totally full and turning people away. Grandma and I watched mom die. In the hours between the respiratory distress and the cardiac infection that COVID-19 would turn to for the death blow, mom told me: "Watch over your grandma, honey. She'll be so lonely."

And she was. The world was too chaotic then for people to do the normal things they used to do when someone died, like gift baskets and housecleaning.

Some people say they remember 2020 as a confusing blur, but I don't. It was just one darkness from January to December, grandma and I pacing around in the 900 square-foot apartment. No one, other than Amazon or Instacart drivers, ever came to see my grandmother, just as nobody has ever come to see me. Eventually things started up again—I wouldn't say they went back to normal. I stayed for a while, watching grandma become more and more isolated as her whole life became an endless tidal cycle of morning television, afternoon shows, in and out, day and night, with nothing to differentiate anything.

Eventually, I went back to school. By then everyone was a year behind, except most of my classmates had spent 2020 learning online while I sat with my quiet grandma. My mom's sister took me in, and I came to see grandma less and less. She died a few years later, alone in her house. I was trying to live my new life with my aunt and I was worried about the normal things a 13-year-old is worried about. I hadn't even texted her in months.

I was rotten for not visiting her, I know that now in my own old age and being alone feels like punishment. Every day stretches out forever, like a musical note held far too long, while the days and years of my life when I had a family and friends spin back further into the darkness of the past.

You're not going so soon, are you?

Robotic Emotional Support and Companionship for the Elderly

"Giving robots more responsibility to care for people in the twilight of their lives may seem like a dystopian prospect, but many see it as an inevitability." (Satariano, Peltier & Kostyukov, 2018)

Some readers might think that the idea of robotic support and companionship for the elderly is much too beyond the pale for any serious consideration—not only is it unrealistic, it is undesirable. After all, who wants grandma to be comforted by the cold metal arms of some human proxy? But let's take a "by the numbers" approach. It is estimated that, by the year 2050, our planet will be populated by 2.1 billion adults aged sixty years or older. Yet, by 2035—a full fifteen years earlier—there will already be a worldwide shortage of about 13 million healthcare professionals, according to the World Health Organization.

As of 2017, there were 46 million seniors (aged 65 and older) living in the United States, comprising 15 percent of the population. By 2050, that population segment will have risen to 22 percent, according to US Census Bureau projections (Anderson & Perrin, 2017). A healthcare shortage awaits. It will be felt acutely by elders with varying degrees of assisted-living needs, many on a 24/7 basis. Waiting until 2034 to start devising healthcare support solutions for the elderly isn't a sound strategy, certainly not in the eyes of any adult on the precipice of joining that demographic.

As in so many other areas of technology, and no doubt partly due to its own "gray tsunami" looming on the horizon, Japan has been way ahead of the curve in recognizing this approaching deficiency. Several decades ago, the Japanese began building and programming robots to help take care of the elderly. One such creation was "Paro," a relational robot resembling a furry seal—a stuffed animal, essentially—that is also branded as a "therapeutic robot" for its positive effects on the sick, the elderly, and the emotionally troubled.

Elder-care professionals have green-lit the presence of these nurturing robots, with cuddly Paro available for patients to hold, pet, and talk to. Seniors and their family members have had a positive reaction to the presence of Paro, finding it "easier to leave elderly parents playing with a robot than staring at a wall or television set." One nursing home patient, Ruth, seventy-two, whose son had broken off contact with her, found Paro a warm presence as a family member surrogate. She would stroke Paro while comforting it and cooing to it, treating the robot as if it's the one that had been abandoned and in need of empathy (Kahn & MacDorman, 2007).

Cynics might say that sticking an aging elder with a robot guided by artificial intelligence is a mismatch, the assumption being that anyone over the age of, say, sixty is as about as comfortable operating an AI-driven device as a time-traveling caveman negotiating the streets of twenty-first century Manhattan via GPS. Or, as Tarantola (2017) writes, "Whether your grandma, who still refers to the microwave as 'the science box,' will be OK living in 1984-like conditions—even if it's for her own good—remains to be seen."

Don't underestimate senior citizens, though, when it comes to high tech. According to Pew Research Center surveys, nearly 60 percent of adults sixty-five and older say that technology has bestowed a mostly positive impact on society, with about 75 percent of internet-using seniors going online on a daily basis. Furthermore, more than double the percentage of seniors who were using smartphones in 2013 are now (at least as of 2017) closely connected to a digital AI world, with four in ten presently owning their own smartphones (Anderson & Perrin, 2017). In addition to the daily web surfing, email use, and smartphone savvy that grandma and grandpa engage in on a daily basis, statistics bear out that many of them are comfortable with or readily adaptable to in-home robot gadgetry designed to assist with basic quality-of-life and healthcare-related issues.

Another robot model (even if not a model robot—it once fell down a flight of stairs during an early demonstration) is Honda's Asimo. Despite its limited mobility, Asimo has shown itself useful in fetching glasses of water and turning off light switches. There's also the Dinsow elder-care robot, courtesy of CT Asia Robotics, which can remind its flesh-and-blood "master" when to take his or her pills. It can also track the user's health and answer telephone calls from family and doctors (Tarantola, 2017). Still other robots or artificial agents have been created specifically for lifting, bathing, or feeding persons in need of such assistance; "conversing" with people and keeping track of personal information; and simulating and triggering empathy through their human interactions (Wachsmuth, 2018). The latter function is key: evidence has shown that elderly people who are lonely or socially isolated are more prone to ailments like cardiovascular disease and infections (Tarantola, 2017).

Not surprisingly, the notion of programmable robots assisting or replacing professionally trained human caregivers has been challenged on ethical grounds, with charges ranging from deception to deprivation of dignity, casting robot users as clueless victims. Think about it (which these robots presumably cannot do): using robots to provide emotional care or companionship to lonely elders could be interpreted as tricking recipients of AI-engineered attention into believing they have a relationship… with an inanimate object. But who's to judge? Unless they are in an advanced stage of cognitive decline, shouldn't we let these seniors decide for themselves what is dignified and what isn't? And even if this ethical objection carried any weight, what is the alternative? Isolation, apparently. Abdi et al. (2018) reference a meta-analysis that revealed that the impact of loneliness and isolation is a mortality risk equivalent to smoking fifteen cigarettes a day. "This is compounded by the fact that social care is a labor-intensive industry in a world with a proportionally shrinking workforce" (ibid.).

The tension between people's superficial understanding of what AI-enabled robots are and can promise to do in the future and the ethical objections that have been raised against using AI-equipped robotic companions to supplement the healthcare workforce are reflected in poll data. A 2012 survey encompassing twenty-seven European Union countries and involving twenty-six thousand respondents found that only 23 percent of those citizens had a negative opinion of robots. The survey's flip side is that more than half of those survey respondents (60 percent) said that robot care for children, the elderly, and the disabled should be banned, presumably for ethical reasons mentioned above (Wachsmuth, 2018).

What we have here is a failure to communicate—a public relations problem only made worse by the many dystopian and malevolent depictions of AI as presented in movies and the popular press. There is on the one hand the demographic reality of an under-served aging population, while on the other there is a perception of using healthcare robots as being either dangerous or nightmarish.

According to Wachsmuth, researchers Amanda Sharkey and Noel Sharkey ("Children, the Elderly, and Interactive Robots," IEEE Robotics & Automation Magazine, March 2011) suggest that interactive robots responding to social cues from cared-for humans might give a false appearance of sentience. This goes back to that perceived deception that appears to stick in the craw of observers privy to elder/robot interactions—

even though granny is benefiting from such interaction, apparently with no strings attached and without complaints from her. When talking about dignity, let's not forget the times that a human caregiver must spoon feed, change a diaper, or pick up an assisted-living facility resident who has been lying helpless on the floor for who knows how long. This isn't just about older or disabled people, either. Kahn and MacDorman (2007) reference an MIT graduate student, who, in discussing the existence of socially assistive robot (SAR) technology, confided that she would just as soon trade in her boyfriend for a "sophisticated Japanese robot," provided such a robot produced the kind of "caring behavior" that would bring a "feeling of civility" to the house.

There also remains the question of authenticity, and whether it really makes a difference to the emotional well-being of a person engaged in some sort of deeply confessional or cathartic expression. Kahn and MacDorman tell of Eliza, a computer program devised in the 1960s to "listen" to users' thoughts and then ask follow-up questions aimed at eliciting responses. The idea was to engage users in a "relationship" that involves the sharing of confidences, as if the human patients were speaking to a human psychotherapist. This is despite Eliza's obvious inability to understand what it was being told or to express genuine empathy for the user. The authors refer to this as a crisis in authenticity: people "communicating" with a computer (robot) "did not care if their life narratives were really understood," Kahn and MacDormand write. "The act of telling them created enough meaning on its own." Perhaps this suggests that a robot that's a superb "listener" can provide a valuable therapeutic service by merely allowing users to get things off their chest.

Robotic comfort toys aren't only for older adults whose various needs require more than just an on-call human caregiver serving an entire floor of facility residents. Similar AI devices are being designed for children beset with chronic health conditions and disabilities that sometimes render human care impractical. Among such child-oriented innovations is the Sproutel-engineered My Special Aflac Duck, which gives young cancer patients a sense of control by putting them in the role of caretaker, "feeding" and "bathing" Ducky via an AR app. Then there's Jerry the Bear, a stuffed animal, likewise developed by Sproutel, which is designed for kids diagnosed with type 1 diabetes, helping them to learn how to monitor blood sugar levels and give insulin shots (Holland, 2018). There's also the Leka smart toy created for children with developmental disabilities, such as autism. It is a small, spherical robot that can be controlled by either a caregiver or the child himself or herself. Its main function: playing educational games with these disabled children, with Leka customizable to whichever level of stimulation and interaction is deemed appropriate for a particular child (Statt, 2017).

All things considered, advancing robotic technology has started to give us choices when it comes to quality-of-life factors, healthcare, and even in dealing with social abandonment or loneliness issues. Left unresolved, abject loneliness even while in an assisted-living facility can be as detrimental to one's health as neglecting to take prescribed medications or not adhering to a healthy diet at breakfast, lunch, or dinner time. Like any other area of science and technology, robot care guided by AI will always be subject to ethical scrutiny, with human dignity and authenticity hanging in the balance. No matter: as human beings, we cling to an innate need to stay connected to whatever it is that defines us as humans, an existence where human relationships and even human touch will forever be needed to make us feel truly alive.

Chapter 7

Alone at the
Water Cooler

The date: Tuesday, November 3rd, 2048.

The place: The 239th floor of the world's first "mile high building," in Tokyo, Japan.

Stefan wakes up, bleary eyed. He'd only gone to bed three hours ago, carousing with a trio of prostitutes (one of them robotic and another that only identified as a quarter female — it was alright, Stefan was a modern man, plus four bottles of sake sure make you lose your inhibitions).

Stefan would have continued sleeping were it not for the multiple, simultaneous notifications he was receiving that he needed to turn in a piece on North American election projections for the next president by noon that day. That gave him around three hours to do the research, fact check it, and then post it to his publisher's queue, where it would be speedily uploaded after having been sanitized of any politically incorrect (intentional or otherwise) verbiage.

Kono, one of his companions from last night, called from his bedroom — "Oh, Stefan? You come back to bed, now?"

Stefan couldn't remember the voice, but didn't care that much... all he wanted was to get a couple more hours of sleep. That's when the idea crossed his mind — JOTBOT 5000 Mk III.

He'd read about it last month in one of his information feeds and it impressed his consciousness for more than the typical microsecond that most stories did these days. Touted as a "fully automated AI writing solution," it promised to be a boon to struggling publishers and news outlets in this day of the "nanospan," the term coined for the new typical attention span of a nanosecond. Rather than spending hours carefully calibrating message, fact, and hyperbole, JOTBOT 5000 Mk III worked on a principle of internet leveraged enhanced deep learning. Using multiple simultaneous feeds that picked up morsels of news data and old public domain fiction from all the world's unguarded servers, and tied in with a highly experimental (although surely robust) algorithm for natural language processing and production using any combination of three persuasive voice models, the JOTBOT promised to do in seconds what authors had previously labored for hours to accomplish. They could puncture the attention span to at least a second, if not more…

Stefan also understood that his deadline was looming, and his boss had already given him two demerits for filing stories late—as his manager never tired of reminding him, "there are a hundred skin bags waiting to take your place, Stefan!… Find the angle, find the story, and pull in clicks and eyeballs, that's IT!!! " One more demerit and Stefan would have to fall back on his universally guaranteed government subsidy, which would put his days of fun with companions like Kono well behind him—he was not prepared to give up this sacred luxury.

With his few hours quickly becoming countable minutes, Stefan jumped on his login to the worldnet and navigated immediately to the host site of JOTBOT 5000 Mk III. He knew his traffic was being monitored by at least five separate levels of surveillance but left the problem of explaining how exactly he was able to turn in his article so quickly to the side for the moment... he knew some of the people in his first and second tier at the news site... he also knew that they liked playing solitaire and creating cat animations more than flagging suspicious activity, so he felt reasonably safe his outgoing traffic would not be scrutinized too much.

The landing screen of JOTBOT was filled with the usual advertising clichés... a quickly changing slide show of single headlines sped past Stefan's eyes: "Eulogies for those EU care about" — "Write it right the first time... aided by AI" — "Jotbot's got lots of hot ideas... Let Jotbot dot the I's for you."

The headlines were accompanied by the appropriate images to allow the visitor to imagine the use cases without too much difficulty... a coffin, a wedding band, and a business-suited executive accompanied each of the headlines respectively, making the target audience clear. What was conspicuously absent from JOTBOT's sales pitch was a headline that read something like: "JOTBOT for JOURNOS — meet your deadlines without anyone knowing." Stefan wondered if his spontaneous headline would make the grade for the site, given the admittedly cheesy cleverness of the ones already present. Also mindful of his ticking time limit, he got to navigating to the enrollment screen.

Like any good web form, the JOTBOT site had the user focus on job requirements first, rather than payment details. Recent marketing psychology findings demonstrated that most people working through the goal of their project first will almost always think of the price as an afterthought, and this was certainly true in Stefan's case. Stefan dutifully populated the fields.

```
User "Stefan" authenticated.
Init_sequence: jotbot.exe
{
TOPIC: The 50th President of the United States.
Target Length: 500 words.
Depth Search (Tier 1 to 8 - Lower tiers are quicker and cost
less, but may provide less than satisfactory results if
expecting writing output at collegiate or higher levels -
suggested min Tier is 3): 3
When does this output need to be delivered by? (Tier 1 for 500
words - expected output 10 minutes; Tier 8 for 500 words -
expected output 2 hours.)
}
```

Stefan was at a crossroads—Tier 1 would give him something fast, to be sure, but would it be distinguishable from kindergarten scribblings with crayon? Tier 8 would take longer but cost more, perhaps even more than what he would get paid for submitting the article. He decided to walk the middle road and requested Tier 5—output ready in 1 hour and 10—this would give him enough time to proofread and send while also keeping a little bit of the money for himself. He put his financial details and submitted the request.

After letting a nice spinner icon gyrate on the screen for the allotted time, and in the meantime grabbing some more under-the-covers time with Kono, the chime on his laptop announced itself. Stefan got to his laptop and couldn't believe his eyes—a perfectly formatted, spelled, and sourced article on the 50th president of the United States. JOTBOT did not need to be told that the election was upcoming and this was an uncertain fact —the tone of the article read like an expert pundit weighing the options. What was even more surprising was JOTBOTs prediction for the outcome of the election for the 50th president— the AI definitely seemed to favor Edna Jamison (an underdog republican third in a pack of ten candidates) over the undisputed "LOCK" for that year's election—William Pearl, a democratic frontrunner who advocated raising the minimum basic income and giving more incentives to non-traditional families. Despite its counterintuitive conclusion, the article was well structured and almost impeccable in its reasoning for its conclusion. Stefan paused to consider whether to submit this article under his byline (and in so doing flout one of the few remaining ethical safeguards in journalism, most of the rest having been erased by the pursuit of the nanospan and the desperation to remain relevant that all professional journalists of the mid-twenty-first century seemed to feel), or ask for an extension and possibly incur another demerit, the third (and most likely final) one he would get before returning to the ranks of the unemployed writer.

The Economics of
Intelligent Automation

"Being worried that automation and AI will hurt society is like being worried in 1440 that the printing press would hurt society and cause mass unemployment of scribes." (James Altucher, Twitter, Jan 1, 2020)

Humanity has long prided itself on its ability to engage in abstract thought and reasoning. Language, mathematics, architecture, engineering, to name but a few disciplines, all have pride of place in our legacy as Homo sapiens. Much of our ability to reason abstractly has led to incredible labor-saving achievements and inventions. Just try to picture the state of the world at its current population without the intricate labyrinth of logistics that connects supplier at point A with consumer at point B. Consider the airplane, automobile, train, and freighter—all transportation devices that have cumula-tively contributed to the staggering success our civilization has obtained in securing its continued survival. Construction machinery like cranes, backhoes, and dump trucks give us similar command over the ground we walk on and develop. Reasoning abstractly has also given us language and the ability to communicate with one another. In order to benefit from the ability to reason abstractly, one must be able to somehow pass along the chains of reasoning and conclusions that one has arrived at through the technology of writing. Recent advances in AI, however, seem to be heralding a new and possibly even more earth shattering eruption into the life of human beings. In fact, this latest wave of innovation could go so far as to make all of our previous efforts at innovation purely redundant, as it aims at making the very process of creative production of text a matter of mechanical contrivance.

The idea of a machine that would be able to express itself about things in a creative way and communicate or at the very least interact with a human interlocutor or audience is an old one. The player piano, patented in 1897 (Encyclopedia Britannica, 2019), is arguably one early example of a machine that seemed to "transcend" the idea of the "machinic," providing entertainment to listeners in the form of music compositions generated by a mechanical piano with a punched roll that served as its musical "program." Charles Babbage was the conceiver of the Difference Engine in the year 1822, and later the Analytical Engine in 1833, the former designed to mechanize arduous calculations such as trigonometric functions and logarithms, while the latter (if built) would have allowed any general calculation to have been entered via a set of punched cards. Alas, Babbage's design never saw the light of day as he had a falling out with the engineer Joseph Clement over the cost of construction (Computer History Museum, 2020).

Contemporary to Babbage's attempts at constructing a machine able to autonomously calculate difficult mathematics given a program as input, there was the Mechanical Turk. Not to be confused with Amazon's crowdsourcing platform for small tasks, the Mechanical Turk appeared to be a machine but actually had a person inside making all the decisions. Its main function was to play chess (or at least, appear to play chess while being controlled from the inside). Famously, the Mechanical Turk beat Benjamin Franklin at a game of chess in 1783, cementing in the popular imagination a vision of what a truly "intelligent automaton" was capable of, even though it was 99% parlor trickery (Standage, 2002).

Fast forward over 200 years later, and it would appear that the example of DeepMind's AlphaZero seems to be the first truly "no strings attached" implementation of a learning algorithm that is able to best its human opponents consistently. In its first iteration, it was able to play chess at a passable level. By its most recent iteration, it is able to beat any chess master that approaches it, without any special adjustments or programming to take into account that player's style. The interesting thing about its algorithm as reported by its creators is that it is a "self-learning" machine (Sadler & Regan, 2019). Rather than laboriously putting in rules, strategies, and mini-max decision points on a game tree up front, the DeepMind algorithm merely plays games against itself many hundreds of thousands of times and then derives the best "policies" from the games it plays against itself using only the most basic ruleset. This is in sharp contrast to the first AI that was able to beat a human chess master in 1997, Deep Blue. IBM's Deep Blue was custom programmed specifically with its opponent Garry Kasparov's games, and after each game a team of engineers would tweak the programming to ensure that it got stronger (Hsu, 2002).

Since Kasparov's loss, it seems there has been no turning back in the era of chess playing entities—engines like Rybka and Stockfish possess chess ratings quantum levels higher than any human grandmaster, and these chess playing "entities" now compete amongst each other to determine which algorithm has chess supremacy. Humans must now look to machines in the realm of learning about chess—we have seemingly taught them everything about this game.

In this continued march towards chess playing perfection, DeepMind's AlphaZero engine became the undisputed world champion entity in the chess world in 2017 after defeating Stockfish 8, which many had considered up until that point the best of the best (Sadler & Regan, 2019). These competitions were no longer about what machine could play better, but were now about how fast a machine could go from absolute ignorance about the rules of chess to full mastery. DeepMind's AlphaZero begins with no knowledge of the game and trains itself to hyper-grandmaster competency in around 4 hours.

After this absolute mastery, DeepMind's engineers turned their gaze towards Go, a previously well-defended bastion of human intuition over machine brute force, human subtlety and pattern recognition over machine rote learning and blistering speed. And it goes without saying that their approach, first applied to chess, eventually went on to prove just as successful in the world of Go. Within a matter of three days and after playing a mere 4.9 million games (!!) against itself, DeepMind was declared the ultimate Go playing entity as well after beating Lee Se Dol, the world's (human) Go champion (ibid.). So much for human intelligence in the world of two player strategy games.

Which brings us back to the possible future fate of human writers and writing. If AI algorithms, using brute force learning techniques and a basic handful of rules, can essentially "bootstrap" expertise out of such starting foundations, unlimited by the constraints of what humans have faced, what should be expected from algorithms that perform the same sort of operations on collections of human text in order to produce reasonable facsimiles of written communication as it is found in the world today, and given the fact that more and more of the world's entire corpus of text, originating from the Babylonians and before, is being digitized and placed online for any bot to conveniently scrape and use as its raw material?

In February of 2019, *Guardian* writer Alex Hern describes in his article a "revolutionary AI system that can write news stories and works of fiction..." (Hern, 2019). The nonprofit behind the innovation, OpenAI, believed at the time that the system (dubbed "GPT2") was "too dangerous" to be released to the open source community, with fears that the text generating algorithm could craft sentences so perfectly that it would be a fake news nightmare. However, within a matter of months, the well-meaning moratorium was lifted, and OpenAI became available to the coding public.

This removal of barriers to entry for would-be deepfake masterminds has led to some amusing results, albeit hardly that convincing to an expert's eye. Hern goes on to give an example of GPT2's handiwork by way of presenting it with the opening lines of George Orwell's *1984*.

In Orwell's opening passage, it reads:

"It was a bright cold day in April, and the clocks were striking 13."

We will leave it as an exercise to the reader to determine which of the two paragraphs below were written by Orwell and which were written by GPT2.

Paragraph 1:

"Winston Smith, his chin nuzzled into his breast in an effort to escape the vile wind, slipped quickly through the glass doors of Victory Mansions, though not quickly enough to prevent a swirl of gritty dust from entering along with him. The hallway smelt of boiled cabbage and old rag mats. At one end of it a coloured poster, too large for indoor display, had been tacked to the wall. It depicted an enormous face, more than a meter wide: the face of a man of about forty-five, with a heavy black mustache and ruggedly handsome features."

Paragraph 2:

"I was in my car on my way to a new job in Seattle. I put the gas in, put the key in, and then I let it run. I just imagined what the day would be like. A hundred years from now. In 2045, I was a teacher in some school in a poor part of rural China. I started with Chinese history and history of science."

Anyone familiar with the story should be able to spot the deepfake immediately. However, someone unfamiliar with the story would be hard pressed to see the difference, or at least to see an obvious "signature" of a machine. Indeed, if anything, the first paragraph (the one written by Orwell in the original novel) seems less "sci-fi" than the second. It mentions Victorian mansions and smells of cabbage. The second paragraph states immediately the "near future" setting. There is, however, something "off" about the second paragraph... something somehow inconsistent... is the setting Seattle or China? Is it now or one hundred years from now? Stylistically, it should be clear that the second paragraph leaves something to be desired that is not present in the first one.

And yet... and yet, what about GPT3? Or GPT20? What will happen when the text generating algorithm is given these "criticisms" and weights its own machine learning accordingly? Could this GPT2, given enough data, write a novel good enough for its author (or more correctly, its programmer) to be paid? What would the ethics of such a troubling picture look like?

Up until now, we've seen examples of AI that could potentially fight our wars and comfort our elders. But what about AI that can produce creative, informative text that is not only indistinguishable from that written by a human, but is indeed even tailor-written to suit each and every pair of human eyeballs that consumes it? While text generation is still in its infancy, means by which an audience can be carefully curated and presented images and information that suits their pre-determined biases already exists at quite a mature level. Ads on the internet target the interests of a human browsing through textual analysis of internet searches a person does (Zuboff, 2019). It is this feature of computer technology that makes companies like Facebook and Google worth so much money. The "downside," of course, is that somebody still has to generate the content that the Facebook/Google algorithms exploit to send browsers along their path of least resistance. But what if that content itself could be generated from previously scraped pieces of writing? Moreover, what if that content was engaging? Compelling enough to get people to click and get advertising to register in their ever shorter attention spans? In such cases, it wouldn't really matter whether the algorithms that generated such content were "conscious" or "sentient." The sponsors of such content generation couldn't care less. Ultimately, what they would be concerned about was whether the process worked to drive engagement and increase sales at the end of the day. And coinciding with this growth of engagement would be an ever increasing sophistication in the types of articles and messages that would end up entering the eyes and minds of future "surfers."

The possibility that these deepfake text generation engines become sophisticated enough to create text indistinguishable from that generated by your average human (journalist/ lawyer/teacher/middle manager/copywriter/content creator ... take your pick) is perhaps one of the most concerning features of the continuing rise of artificial intelligence research and applications. Given an even moderate ability to generate text compelling enough for consumers to read, and the cost savings quickly add up. One could conclude that such AI engines will take over an increasingly large percentage of the work that is currently done by vast swathes of the "middle class." Where the industrial revolution of the late nineteenth and early twentieth century automated and routinized physical labor, thereby displacing thousands of factory workers, so the current AI revolution seems to be promising a similar displacement of a number of "white collar" jobs.

On the other hand, however, one must return in considering such questions to exactly what it is in writing that engages and compels people to read more. Would a short story, created entirely by an AI, truly be able to be just as compelling as a short story written by the likes of, say, Alice Munro? At this point in history it seems doubtful. Here, for example, is a passage from one of Alice Munro's more famous short stories written near the end of her career and at the height of her powers, "Dear Life," where she describes what becomes of a childhood friend:

> "I ran into the grandmother now and again. She always had a little crinkly smile for me. She said it was wonderful that I kept going to school, and she reported on Diane, who also continued for a notable time, wherever she was—though not for as long as I did."

So much contained in this brief paragraph. What AI would describe a smile as "crinkly" without seeing and knowing a human face? What about Diane? Why would her time in school be thought of as notable, except in a case where small-town gossip made your business the business of everyone else as well, a fact that would be lost on an AI. For comparison purposes, one of the paragraphs below was generated by Adam King's implementation of GPT2, "Transformer," found at https://talktotransformer.com. The other is Munro's continuation (Munro, 2012). Try to spot the deepfake.

Paragraph 1:

> "It was the social anxiety I went through that eventually afflicted me with crippling emotional insecurity, which is reflected in my giving up my house to depend on no one else for support. I didn't know at the time that this was also true of my addiction to alcohol and drugs."

Paragraph 2:

> "According to her grandmother, she then got a job in a restaurant in Toronto, where she wore an outfit with sequins on it. I was old enough at that point, and mean

enough, to assume that it was a place where you also took the sequin outfit off."

Without being familiar with Munro's story, there again seems to be superficially little difference between these two passages. However, at closer glance, there is something disconnected and almost jarring about the first paragraph—there is no mention or indication of social anxiety or addiction in the initial paragraph; nor is there any mention of a "house" that must be "given up." And logically, why would giving up a house allow one to become more independent, as implied by the first paragraph? For now, at least, it appears as though GPT2 is unable to replicate the intricacy of Munro's prose. Looking ahead, however, and given the current state of the art in text generation, there are three general outcomes that can be considered:

First: text generation will continue to get better until eventually engines like GPT2's descendants will be able to craft compelling text that will be not only indistinguishable from writing like Alice Munro's but will even compete with writers for accolades and prizes.

Second: text generation will never be able to match the nuance and subtlety of a writer like Munro—human writers will always have the edge on such creations.

Third: text generation by machines will not be able to compete in quality with that of human beings, but it will not matter—humans at large will become so used to text generated by AI engines that they will frequently skim over much writing and not worry too much about how it was generated.

The possibility of (1) being true still seems so remote that this writer feels it is completely out of the running. Possibility (2) sounds far too like bad predictions of the past about what machines would "NEVER" be able to do. It is the possibility of

(3) that is really very alarming, and it follows a familiar path in the way that many other technological inventions and innovations have set the precedent for. It goes something like this:

We used to use horses and horse drawn carriages for traveling long distances. This method of transport had the advantage of being able to take the rider in almost any direction—a horse is able to navigate many sorts of terrain and can therefore allow relatively remote populations contact and access with others while also giving them privacy and autonomy. However, over time, horses tend to make tracks in the landscape, and the well-traveled routes become part of the everyday paths people take when moving from point A to point B. When the car comes along, the well-traveled paths become the beginning of roads, and as cars are not nearly as flexible as horses in terms of getting from place to place, the paths need to be paved. Once the paths are paved, cars get people places faster, but there are fewer places to travel, and so populations tend to grow only where cars can connect the nodes. Meanwhile, horses fall out of favor because they cannot use the roads that are now built for cars due to safety hazards, not to mention that they now need special shoes to walk on the newly paved surfaces. Thus, the world becomes transformed, not in order to enhance the lives of the people in it, but instead to enhance the usefulness of the new innovation that promises so much if only the world conformed to its design.

The same phenomenon could occur for text generation—short stories written by human beings will become like the small villages and homesteads that used to dot the landscape. Easily reachable by horse but requiring a not insignificant investment of time, the villages and homesteads would be unique and charming. As text generation grows, the incentive to produce any original text is slowly sapped, as "roads" of text are formed between various sources, and the overall appetite for the genuine written word declines as machine generated "content" becomes cheaper and seemingly just as good as the human created counterparts. Finally, the infrastructure of machine generated text will eventually become so entrenched that authors will need to learn how to become "content producers" and ensure that they are following the means by which machines are generating and distributing content. Rather than spend the effort to try to write something authentic and reflective of the human condition, future authors will instead choose to gather the "low hanging fruit," exploiting short attention spans and emotional reactions to inflammatory headlines in the same way that horses exploit shoes in order to walk on paved roads. Ultimately, however, the genuine authors and journalists, much like horses today, will become rare, special purpose animals that exist for exhibition purposes more than practical work.

Whether one thinks that the scenario outlined above is "dystopian" or "utopian" is ultimately a matter of esthetics and value judgment—the authors themselves are somewhat divided on what this future might mean from an ethical perspective. From a purely practical perspective, it seems naïve to think that the genie of text generation can ever be put back in his bottle. What has remained true of humans in our history continues to obtain—we like our technology, and we continue to improve it for as long as it remains viable and until it gets replaced by something else. On the dark side, we might bemoan the fate of journalists and creative writers and argue that there is simply one more job being taken by "the robots." However, keep in mind the quote at the beginning of this chapter—this may be similar to how a person might have reacted when the printing press was introduced—wondering what all of the scribes were going to do with their free time.

Chapter 8

Do As You Are Told

"If we built an AI, I think we can agree it wouldn't naturally have a desire to sit on the couch and stuff its face, become an attention whore, or have sex with hot young people. It wouldn't want to do those things unless you designed it to want to do those things. By the same token, I don't think an AI would want to dominate, discover, or even survive unless you made it want those things." (Young, nd)

In the movie franchise *The Terminator*, an artificial intelligence by the name of Skynet (or Legion, in an alternate timeline) becomes self-aware and decides in a nanosecond that humanity is a threat to its existence, thereby leading it to launch a mutually destructive nuclear exchange between the (human) world powers. AI doomsayers have used something like this scenario to warn against the continued pursuit of AI, at least without examining the safeguards that would be necessary to prevent a "Skynet" from happening in real life. Descending from the heights of the fictional world to this real one, however, let's take a look at what would have to be true, logically, philosophically, in order for a Skynet-type scenario to really become a threat, i.e. that an actual machine intelligence was "born" and in a moment "decided" to launch some attacks on humans to ensure that they could eliminate their "competition" and continue on along without any biological assistance at all.

We will argue that there are three things that would necessarily have to be present in order for Skynet to "become sentient" and make decisions that would intentionally aim at the destruction of life, and two of them are currently completely out of the running as possible real futures due to the fact that we have no idea of what it would take to make them true. This is because we currently do not have a theory that manages to account for (let alone describe how we might be able to transfer) the qualities of "will" and "desire" into any of our artifacts. For those who may argue that "will" and "desire" will eventually prove to be "folk conceptions" of much more trivial and completely physical descriptions of consciousness (the physical eliminativist's holy grail espoused by the likes of the Churchlands and Dennett), the ball is in their court to demonstrate how such things as will/desire would ever be embodied in a machine. Nevertheless, it will also be argued that we cannot trust those in power to make the correct decisions regarding what sorts of models AI agents should embody for the sake of the public good. And while it will be argued that, in principle, the creation of a sentient AI is impossible due to the inability to have a machine autonomously instantiate will or desire, it is nevertheless POSSIBLE to create an insentient AI that makes people's lives hell in practice, due to the ability of certain key people in powerful positions deciding what both WILL and TRIUMPH will mean in a value hierarchy where AI is omnipresent, omniscient, and omnipotent.

In the meantime, we can examine some initial approaches and then analyze them to understand why they must fail to provide the machine with any sort of desire to remain alive or instinct for self-preservation, as well as any sort of way of generating and then pursuing self-directed goals of behavior and achievement.

Any machine that wants to become Skynet will need three things: 1) an ability to make decisions and act on them based on multiple views of data; 2) a will to consistently improve its own position in terms of executing these decisions; and 3) the desire to ultimately triumph (whatever triumph may mean to the sentient entity, this will become one of the major obstacles, as we will see) in the conceived field of possibilities that such an entity might possess, i.e. there would never be any such thing as a "win-win" scenario involving co-operation for such an entity, as its desire would be for its sole victory alone.

Taken in turn, we can see right away that the first pre-requisite is relatively easy to meet, at least in a conceptual (and increasingly, practical) sense. As we have seen in other chapters, many machine learning and artificial general intelligence techniques are all about providing decision trees and weighted alternatives based off of various data given at differing levels of complexity and analysis. AlphaGo Zero is successful at playing Go because it has been given the ability to create move policies from an inconceivably vast set of previously played games. Moreover, there is a straightforward method that exists within the rules of the game of Go that allows it to assign "better" and "worse" as descriptions of each move that has been or is going to be played. Thus, for computers at least, the first pre-requisite is the bread and butter of machine intelligence, operating well within the wheelhouse of what makes computers in general, and artificial intelligence more specifically, so formidable.

However, the second pre-requisite poses some more intriguing, if not intractable, conceptual challenges: how can we instruct a machine to consistently improve its own position within more abstract hierarchies, such as let's say human society? Can we think of any way to specify with enough detail the generalized "will" of what allows us humans to strive on a daily basis towards all different manner of goals that we might formulate? Let's take AlphaGo Zero again as an example. Certainly it is able to generate better and better ways of playing the game of Go with each iteration of gameplay… but we have told it to do this. The machine has not spontaneously decided to pursue the game of Go on its own—writing these words should indicate to you the difficulty (if not absurdity) of even trying to get a machine to do something on its own, without being told (explicitly or implicitly) what it needs to accomplish. Compare the AlphaGo Zero situation with a human who decides to start playing Go. AlphaGo Zero has been compelled to play Go—when electric power is sent through the various circuits and semi-conductors of whatever hardware AlphaGo Zero runs on, in whatever patterns and branches that are determined dynamically by AlphaGo Zero's machine learning algorithms, it merely executes the next step in its already defined mission. A human who decides to play Go will be coming at this activity from a very different angle, however. Perhaps the human will have wanted to play Go because they like board games in general. Possibly a member of their family used to play the game and they want to continue the tradition. Maybe they would just like a new hobby to pass time. At every juncture in the pursuit of playing Go, the human will have a host of other activities that they will also be able to pursue— maybe there is no time to play Go because the human needs to pick up their child from daycare. It might be that they begin to play Go but become discouraged shortly after due to being beaten badly by someone (perhaps AlphaGo Zero!) and deciding it is not worth the energy or effort, because there is no hope of becoming as good at it. At each step in deciding to play (and persisting in playing) the game, there are a host of competing priorities that the human subject must contend with. When the human is able to choose the particular pursuit of playing Go over and above other pursuits, we can say that he has a "will" to become better at Go as opposed to some other pursuits. Now, notice that we can dispense with the thorny question of whether this hypothetical Go playing human has "free will." It matters little for the present argument as to whether the human "freely" chooses one pursuit over another. (Indeed, without going into this rabbit hole too deeply, one can ask at this point what this could even mean.) What is salient here is that there are a whole host of other preoccupations and projects that the human intelligence will have available to them, and it will be their "WILL" that determines which of these many alternatives will be executed (Schopenhauer, 1969).

Now, the AI evangelist/transhumanist may respond at this point by saying: "Yes, of course AlphaGo Zero only has parameters to play Go but, as you have described it, one could easily say that the moves AlphaGo Zero makes all come from a similar form of deliberation as a human uses, albeit more focused on a particular goal. We can merely add the number of tasks that some general intelligence could 'choose from' and again let the machine learning algorithm iteratively chug along until it came to some decision."

Let's try this out with another thought experiment. Assume we have constructed an all-purpose abstract game-playing entity that does not just play Go or chess, but can play any of a whole host of different turn-based two player games with perfect information—call it AlphaOmni for the sake of this exercise. We will not only allow it to learn all such two player games, but we will even try to give it some choice in which two player game it "wants" to play. Here comes the conceptual problem—how will we tell AlphaOmni how to decide which game to play? When it is a specific game, the issue does not arise—it is easy to define the victory conditions in a two player zero-sum game because that is exactly what a two player zero-sum game is all about—reaching a state of "win" given a set of board positions and inputs for movement. But how does one win given the choice of whether to play chess or Go? What about Othello versus checkers? How would you decide? Perhaps you might choose checkers over chess in a proposed game with your grandfather, as you have fond memories of playing this with him as a child. Perhaps you would play chess with a friend, as you have a chess rivalry that goes back many years and you have a score to settle. Both of these motives rely on a functioning system of emotion and volition—a "felt sense of meaning" that serves as the motivator for an intelligent agent to make a selection from an open ended set of options. Indeed, in order to put the "G" in the AGI for our hypothetical agent, it is simply not enough to have it decide to pursue one option from a set of many of a similar class, but it would have to pick from an open ended and arbitrary list of options. Due to our biological embodiedness (refer to "The Importance of Embodied Human Intelligence" chapter), there are a set of constraints we must constantly strive to fulfill—nutrition, hydration, rest, companionship, etc. These constraints in many respects inform what we decide to do next, but not always… a monk might decide to meditate in a cave for days without food. What is the motivation for this agent to perform this action? Would an AGI have to simulate this type of action as well in order to be considered truly intelligent and self-directed in the same way humans are? We would argue yes, they would, and since we still have no theory of mind with which to explain how or why human beings do what they do, we should not expect that machines can be magically conjured to do so either, at least not without ultimately having the decision preferences of its creators inserted instead—this is an important point we will return to later.

The third element of AGI that would seem to be both necessary and currently intractable in order to produce actual goal-directed behavior is the ability to triumph, or at least move in a direction that is "better" than the current state in which an agent finds itself. Again, we emphasize that this is a general feature of intelligent behavior in all animals (embodied biological intelligences), and it is not enough to object that machines can be given arbitrary goals towards which they may evolve. Clearly, AlphaGo has been given some idea of moving towards a "victory"—if not, it would not play Go with any sense of improvement at all. However, it is relatively easy to think of how to define things like "progress," "better," and "triumph" in such a zero-sum environment. The same cannot be said for the dizzyingly complex world we find ourselves navigating every day. Think of the kinds of decisions you have had to make today—this past week—this past year. On a daily basis: should I pack a lunch or eat at a restaurant? If I optimize for cost, perhaps bringing my lunch is the best choice, unless a restaurant is offering a special today and I have a chance to eat with some friends. If I optimize for nutrition, perhaps eating at that vegetarian restaurant is best—but what about that time I ate there and didn't feel well afterwards? And how do I decide what to optimize for?

The reason humans are able to break out of this seemingly infinite regress of considerations is that we have deeper constraints of time and biological viability that, in a sense, force us to make the choice. There is, in the final analysis, always a hint of arbitrariness to our decisions. However, they are not merely random (except when we decide to make them so, perhaps by flipping a coin to decide which restaurant to go to —but even deciding to make the decision random is itself a higher-order decision that we really have no insight into how we make). It is unclear at this point how an AGI would be "trained" (or even "self-trained") to make such arbitrary decisions guided by both reasons and feelings, and the argument in this chapter is that it is in principle impossible to generate or discover an algorithm that would be able to make such arbitrary decisions with the same level of autonomy and finesse as humans currently do.

The current state of the art in machine learning (the method by which artificially intelligent entities such as AlphaGo "learn" to play games) consists of an array of mathematical and statistical techniques that are well understood but difficult to iterate through computationally using brain power alone. This is where the strength of computers shines—brute force processing using a training set or set of constraints it must fulfill. There is, and should be, nothing that suggests to the casual observer once they understand this basic fact that AlphaGo "understands" what it does, or has any sort of sentience or awareness of what it is doing. It is merely a very sophisticated calculator. And until such an all-purpose algorithm is produced, the goal of actually replicating sentience in our AI endeavors remains out of reach.

At this point, assuming the reader is sympathetic to the real possibility of genuine AGI and has made it this far without throwing the book across the room, we can entertain some of the typical objections that the authors of this book have received and vetted given this line of argument.

Objection 1: what you call "will" I call the certain and indubitable mechanistic/computational responses of an agent that has been programmed appropriately or correctly (depending on context). "Will" in the folk psychological context of the metaphysics of intentionality is little more than a placeholder for things we do not understand. Once we iterate through a number of generations of producing or programming such intelligent agents, it will be clear that the question of "will" was never so important, after all.

Response: this objection will typically be given by theorists who are approaching the problem of AI from a very high conceptual level and not from the practical, "in the trenches" viewpoint of an applied computer engineer — I'm talking here about folks that have as their daily bread the uses and limitations of computers, and navigate them all day, every day in consultations with various business stakeholders to discuss what is and is not possible given a) the current state of technology and b) the current state of their already existing IT systems that would enable them to leverage the larger current state of the art. The question from the foregoing is re-asserted: specify to me an algorithm that would be able to settle upon a course of action that is drawn from a completely arbitrary and unstructured list. Here are the things I have thought about doing in the past half hour: a) go to the bathroom; b) go out to buy groceries; c) make dinner; d) write this book; e) plan a vacation; f) send an email; g) research the Sicilian defense in chess; h), i)... Taken one at a time, sure, an AI would be able to solve for it. But taken as a whole? It is very reminiscent of Borge's Chinese encyclopedia, categorizing the 14 known classes of animals (Borges, 1964). It reads as follows: (a) belonging to the Emperor, (b) embalmed, (c) tame, (d) suckling pigs, (e) sirens, (f) fabulous, (g) stray dogs,

(h) included in the present classification, (i) frenzied, (j) innumerable, (k) drawn with a very fine camelhair brush, (l) et cetera, (m) having just broken the water pitcher, (n) that from a long way off look like flies.

What structured algorithm would be able to take into account all of these various possibilities? Answering that the "computer will learn for itself" will do no good, as evidenced (and described above) by the constraints one must place on a learning algorithm before it even begins to become "intelligent."

Objection 2: "desire" is a placeholder concept left over from a misguided intentional stance towards mind that can easily be accounted for in terms of more modern vocabulary as "optimization" — if I give a sophisticated AI algorithm the parameters that need to be optimized iteratively, I can easily achieve something like a simulation of "desire."

Response: as mentioned above, it is the decision of which parameters that need optimizing (and are sensitive to contextual inputs) that is the hallmark of intelligent behavior in embodied organisms, and not the insertion of optimization criteria per se, that is the key issue that is being highlighted here. In any given novel problem, an agent must determine a number of "meta-problematic" factors: a) how important is this problem to be solved?; b) how much time does one have to solve this problem?; c) how many different solutions can be explored, i.e. what is the resource cost of exploring each and every solution?; d) how important is it that each and every alternative be explored?; e) how much net benefit will any one of the solutions provide, either to the agent or to the stakeholders for which the problem is concerned? In the case of the AI examples we have already discussed (chess, Go), filling in these blanks is straightforward. Similarly for the limited areas in which things like expert diagnosis in medicine and precedent search in law are proving very robust. Taken more generally, however, it is inconceivable how ANY AI algorithm would be able to solve for these constraints autonomously.

Objection 3: if we grant that no AI can solve for either "will" or "desire" as currently understood, at the very least their creators can put in "placeholder values" by hand until such time as a way of solving for these variables becomes more practicable.

Response: this is entirely possible, and there is no straightforward refutation of this case. Human creators CAN (and indeed DO) put implicit aims or constraints into their algorithms. But note now that what was once a purely scientific problem has now become one with political dimensions. WHO is embedding the constraints that the AI is to solve for? WHAT INTERESTS does this activity serve? And WHAT is the eventual outcome of such AI agents making their "autonomous" decisions?

It is not that the Terminator scenario outlined at the beginning of this chapter is so far-fetched... At least, it is only far-fetched insofar as the machine would arrive at its own value judgments. Rather, what makes the Skynet scenario so paramount to our actions in the next decade or two is that value judgments could be smuggled into the operations of an AI without anyone noticing, at least not at the outset. By the time such decisions seemed to go against the will of the people for which they would serve, the state of AI today can make no comment, as the decision algorithms or "policies" that AI creates are so beyond the casual (or even deep) inspection of their creators that they begin to become little more than "black boxes" into which problems are fed and out of which solutions are generated. And it is the nature of human bureaucracy to easily adopt policies that seem to promise short-term gains while in the long-run promise to provide headache and consternation.

The pursuit of AI as a human endeavor promises much. The ability of machines to survey landscapes of thought in small fractions of time compared to how long it would take for a human to do the same will no doubt become a key feature of all human research in the next one hundred years. The deeper concerns only arise when we are lulled into a false sense of belief that these machines are behaving autonomously. It has been shown above that this is impossible, or at least currently intractable given the state of the art in AI research. Rather, machines are and have always been things that "do as they are told." And it is up to all of us to ensure they are being told to do the right things, whatever these may be.

Chapter 9

The Devil Made Me Do It

The defendant was one Harold J. Pritch, of New Vesebius, of the state of Ormaton, a small minopolis with approximately three million residents.

The charge: vehicular manslaughter.

Details of the Event:

Mr. Pritch was being driven one night towards his home in the hills of New Vesebius. More specifically, his Ultron9000 electronic conveyance was taking him to his pre-agreed destination—339 Orangemount Drive, a pleasant little hilltop bungalow that looked out over the tiny village. During this drive and coinciding with Ultra9000's route was a night cyclist moving in the same direction. Due to the orientation of the "self-driven" vehicle's sensors combined with the relative motion of both car and bicycle, the presence of the cyclist was not registered by vehicle, so no evasive action was taken as the car passed the bicycle.

The cyclist, Robert Mornington, was knocked from his bicycle after the Ultra9000's front right fender collided with his rear wheel at a relative speed of around 30 miles per hour. Mornington was ejected from his bike and fell to the left, where he was quickly pulled under the wheels of the vehicle. An emergency collision alarm was activated and the Ultra9000 came to a quick (but safe, non-skidding stop, as it had been programmed) while Mornington was under the vehicle, causing his body to be dragged and rolled along for the full 29 meters it took to stop. Mornington was pronounced dead at the scene.

Mr. Pritch was questioned by police and charged with vehicular manslaughter. The prosecution claimed that, as the mandatory human occupant in a self-driving car, Pritch needed to ensure that he was monitoring what the vehicle was doing at all times, as per state and federal legislation. The defense argued that Pritch did indeed see the cyclist but the safety override that Pritch attempted to activate did not work in enough time to prevent the accident. Pritch was acquitted of all charges. Edison Inc., the company that makes the Ultra9000, has been sued by Mornington's family in a civil action. Edison Inc. spokeswoman Deirdre Hall has gone on record claiming that the company is not liable for the death for the same reason that all car manufacturers are not liable for accidents with their automobiles — they cannot be held responsible for the use or misuse of their products.

Out of Order

> "The right understanding of any matter and a misunderstanding of the same matter do not wholly exclude each other." — Franz Kafka, *The Trial*

These are the words spoken by a priest in Franz Kafka's dystopian nightmare *The Trial*. The main character (simply known as Josef K.) is informed enigmatically at the beginning of the novel that he has been arrested for some crime, the details of which are never revealed. Told to report to a courtroom on an upcoming day, K. is not given any details as to where or when his "trial" is meant to take place. Finding his court appointment almost by accident, Josef K. is slowly drawn into a world of ever greater absurdity, having to overcome one after another meaningless bureaucratic obstacle. The priest's utterance is a mysterious and paradoxical pronouncement on the nature of the irreducibly human concepts like guilt, responsibility, and justice. *The Trial* as a whole is an allegorical depiction of the opaque workings of the justice system, and indeed the paradoxes that it puts people through in its attempt at "serving justice." In a world of AI and monolithic mega-corporations that possess and control the data of hundreds of millions of human beings, Kafka's nightmarish depiction of an individual caught in a system he neither understands nor belongs to is prescient. The justice system has never been known for its ability to quickly innovate new solutions and become more efficient as case loads increase. On the contrary, it is in the very nature of the justice system to operate slowly and deliberately—the outcomes of decisions made on major cases within the justice system can frequently have life-altering impacts to all the parties involved. For minor disputes, case backlogs and inefficiencies dissuade people from pursuing remediation, an ultimate failure of justice.

Over the next few decades, technological changes (not the least of which is the introduction of AI at all levels of industry and society) are poised to exacerbate this already unfavorable situation. A future world of increased artificial intelligence, massive information surveillance and storage, and an exponential increase in network traffic and transaction volume will pose a heretofore unimaginable burden on the judicial systems of all countries and will require some form of technological innovation to mitigate. Furthermore, despite the risks, failure to innovate in the historically innovation-resistant area of law and judicial due process will lead to ever increasing backlogs of casework and ever increasing miscarriages of justice. Settling disputes justly and in a timely and efficient way will become increasingly problematic in our disruptive and technologically dominated future. Artificial intelligence and crowdsourced solutions may help, but it is essential they are introduced judiciously in order to ensure that Kafka's novel does not become prophecy.

For as long as there have been human beings, there have been disputes among human beings and the need to resolve such disputes in the most harmonious way possible. It is the role of the law and the court system to provide just such resolutions. Brenner (2007) usefully distinguishes between "civil" and "criminal" law:

> "Criminal law differs from civil law in two very important respects: One is that a criminal case is brought by the sovereign (a state or federal government in the United States), while a civil case is brought by a private person."
> (p. 7)

In criminal cases, it is the "sovereign" who prosecutes the case — in essence, the government brings action against some party accused of criminal wrongdoing and formulates a punishment. Civil cases are brought by complainants in an effort to have their damages they claim mitigated or repaired.

We know that a formalized means of resolving disputes was memorialized some 2000 years ago in Ancient Mesopotamia with the famous "Code of Hammurabi." By way of externalizing the eternality and rigidness of the code, it was etched in Babylonian script on a hard glass-like slab. Much like the imposing and inflexible Code of Hammurabi, the modern legal system is meant to be stable and slow to change. For many centuries, this was not only a natural feature of the justice system, it was *desirable*. When technology did come along to disrupt traditional patterns of legal judgment, it came along in discrete and manageable packets.

For example, the rise of rail transportation brought with it a host of legal challenges and implications. Things like deciding how much passengers and industry would pay, creating and enforcing safety provisions, and granting bonds to raise capital for continued development were all "context-specific" features of the railroad industry itself (Brenner, 2007, p. 10). Other industries that have relied on sharply delineated contexts of technological advance have similarly posed few problems to the legal system, including things like airlines, machines that are used in a manufacturing or primary resource extraction capacity, and even the realm of cinema (ibid., pp. 25–74).

The same cannot be said for the internet—itself a technology that from inception has been resistant to various forms of legislation, dispute resolution, and enforcement. At the center of much law that governs computers and the internet is the concept of "use" and "misuse," with criminal acts involving the internet centering around some implicit or explicit concept of "misusing" the resource. However, due to the inherent flexibility of computers when joined to the communicational backbone of the internet, there is no "context-specific" way in which laws can be drafted for these technologies, and today's "misuse" can very easily become tomorrow's "innovation." As Brenner writes: "...the 'misuse' of computer technology has received a great deal of attention from legislators over the last two decades. Much of this legislation has been reactive—a legislature's responding to an actual or perceived 'new' threat such as fraud or phishing" (ibid., p. 117). This has resulted in laws that are redundant and difficult to apply specifically to the broad technological base that all forms of connected computing devices comprise.

Laws that pertain specifically to the "use" or "misuse" of a computer in the pursuit of a crime clearly have in mind a "user," and this is certainly one of the areas in which artificial intelligence will pose issues for this sort of legislation. Looking ahead without taxing the imagination or reality too much: who is the "user" of an automated vehicle that has no passenger? How about the user of an automated classification system used by hospitals for billing? Or the user of an automated cleaning robot that patrols corridors of an office building in the middle of the night? One does not even have to foray into the exotic world of AI to see that the technological space of data management and integrity is already highly complex and a virtual nightmare for anyone who happens to get compromised by it.

The growth in disputes occurring specifically in the virtual world has been marked since the mid-nineties, when the internet began its explosive and intrusive growth into the lives of everyone. In describing this growth and the concomitant rise in disputes, Katsh & Rabinovitch-Einy (2017) write:

> "Our activities online and offline are taking place in an environment that is active, reactive, and, for some, lucrative; it is not, however, friction-free and harmonious. In any environment, the more relationships that are formed and the more transactions that take place, the more disputes are likely to occur." (p. 3)

Many changes have been wrought in the technological landscape as a result of the proliferation of the internet, not to mention the rise of artificially intelligent algorithms that gather and store data about the network's users and habits. Database hacking, identity theft, and the unauthorized use of personal information by agents of companies that remain utterly unknown to the victim are becoming more commonplace. Now arguably for the first time in history you can become a victim of a crime without even knowing it. Indeed, looking again at Katsh and Rabinovitch-Einy's treatment, they define a new crime: "identity pollution" — what happens to your digital identity when aspects of you are miscategorized, over-generalized, or otherwise mistaken about who you are (ibid., p. 5).

This can be seen perhaps most dramatically in the case of Dave deBronkart, a now 70-year-old cancer survivor who experienced a modern day Kafkaesque nightmare journey through the computer-mediated medical landscape from 2007 to 2009. Beginning in 2007, deBronkart was diagnosed with stage 4 kidney cancer — a disease with a grim prognosis in most cases. Using his computer background, deBronkart did his best to ensure he used all the technological tools at his disposal to defeat his disease. Taking to the internet as "e-Patient Dave," deBronkart blogged about his experiences taking front and center accountability of his healthcare.

And it was a good thing he did, too. DeBronkart elected to use a patient data transfer system that would allow his local hospital records to be uploaded to (the now defunct) Google Health to facilitate ease of sharing with other healthcare providers in the form of a "personal health record" (PHR). While the process sounds ideal, in practice it was anything but — deBronkart's transferred data not only failed to contain whole sections of relevant information, such as lab and radiology results... it also *did* contain a host of inaccurate information, such as reports that his cancer had spread, a contraindication for an irrelevant medication, and a list of conditions he had never had (DeBronkart, 2009).

Given DeBronkart's somewhat horrific experiences above, two things are worth mentioning. If DeBronkart was not as vigilant or involved in his healthcare, we would never have known anything was wrong with his file. If the system was not experimental but very well entrenched, we can only imagine how much more effort DeBronkart would have had to exert in order to correct his medical file. Nevertheless, the current laws in place for computer crimes are unable to fully account for DeBronkart's torment under theories of "use" or "misuse." As Katsh and Rabinovitch-Einy (2017) write:

> "Judicial decisions attempt to clarify what the legal standard is and communicate to the public what is allowed and what is not, all with the hope that these rulings will be followed and future disputes averted... While the conscious flaunting of legal standards is the cause of some disputes, unfortunately most are simply the consequence of inter-actions gone badly, of bad data being employed, or of good data being used badly." (p. 16)

As the volume of data continues to increase (whether it be medical, financial, demographic, etc.), it is inevitable that the volume of disputes surrounding incorrect data will also increase. Even with sophisticated AI learning tools, bad data can never be turned into good data as the AI has no way of telling the difference. What's worse, most methods by which current machine learning algorithms arrive at decisions are more or less opaque to their users. While an engineer might be able to understand the process by which deep learning neural networks change their weights, they have no insight into why or how a particular learning algorithm has done so. Given the rapid pace of technological innovation, there is a huge land-scape of potential human misery that current legal systems have only a limited ability to deal with. What are some potential solutions?

Two innovations that some large internet companies have begun to adopt are alternative dispute resolution (ADR) and online dispute resolution (ODR). ADR is a means of resolving disputes without having to go to court or be involved in a lengthy and costly legal filing. Coming to prominence in the late 60s and 70s, ADR consists of four separate dispute resolution processes: negotiation, mediation, arbitration, and collaborative law. ADR itself came out of the fact that traditional cases put before courts were taking longer (and therefore becoming more expensive) to settle, making the very hope of achieving some sort of justice for both parties in the dispute counterweighted by considerations of time and expense. Clearly, when the ten thousand dollar case being prosecuted on is on a timeline of 5–10 years and will cost over one hundred thousand dollars to complete, it is obvious that the courts' mandate to dispense justice is no longer being upheld. It is in this vacuum left by the courts that ADR (and increasingly, ODR) fills a gap.

Given the volume of online transactions that many internet companies process on a daily basis, it is not surprising that the seemingly small proportion of them that end up being disputed is massive. Take, for example, AirBNB. AirBNB is a popular "accommodation sharing" site that allows users to either "host" or "guest" in properties throughout the world. For travelers who would rather not stay in an expensive, institutional hotel, the appeal is that they will be able to take homier accommodations, and frequently at cheaper rates than they would otherwise pay. For hosts, the site provides a means by which they are able to make more money from their property or extra space without having to go through the hassles of licensing or capital investment that becoming a hotelier would require. While this sounds like an ideal "win-win" situation for both guest and host, the devil is in the details. From the guest perspective, there are numerous cases where the host is unavailable for the all-important check-in and handover of keys for multiple day stays (Collinson, 2018). Or the host has multiple listings for the same property, and is attempting to scam people by having multiple parties book at the same time (Temperton, 2020). Or a host provides sub-par accommodations but deceives by posting fake glowing reviews of his own properties (Dubminsky, Ouellet & Forero, 2019).

And hosts on the site do not fare much better when it comes to scammers. Countless examples can be found where mansions, homes, and apartments are trashed by careless and malicious guests (Mills, 2019; Padojino, 2019; Bontke, 2019). Given the fact that AirBNB processes literally hundreds of thousands of transactions a year, it is not surprising that recently the company filed a patent for a "trait analyser" that would scan various aspects of a person's online identity (think social media presence and news stories) in a bid to prevent accepting reservations from such "undesirables" (Bourne, 2020).

While the move to innovate such a solution is under-standable in light of all the potential disputes, a number of troubling questions also crop up with respect to using technologies to arbitrate what are, at base, human disputes stemming from human characteristics. For instance, will the application have a built in "statute of limitations" beyond which it will not consider? Will it have its own database of information that it compiles and sells back to the hotel industry, perhaps creating a sub-class of "unrentables," citizens who are no longer trusted with the ability to either be host or guest in rented accommodations? The parable of Dave DeBronkart is once again instructive here—through no fault of his own, a computer system decided he was a sicker man than he or the doctors believed. If it weren't for the fact that DeBronkart was on top of his digital identity, who is to say what the ultimate damage would have been to him and his family due to the "identity pollution" he endured? In cases of "unrentable" hosts or guests, who knows what subset of these individuals would be savvy enough to investigate their own online fingerprint, searching for "false positives" reported to information gathering agencies and bots to determine their own rental worthiness?

Internet site eBay is another business that has grown incredibly over the past decade, necessitating its own "in-house" forms of dispute resolution. A major problem since eBay's inception was managing identity and reputation pollution. Sellers who received one or two negative pieces of feedback would always have them on their profile—there was no "judicial recourse" for redemption or second chances. In 2008, it introduced a "community court" in which "sellers disputing negative buyer feedback could submit their complaint to a randomly selected panel of jurors. This represented one of the first attempts to use crowdsourcing in dispute resolution. It also allowed eBay to address the challenge of handling the relative small percentage but still significant number of disputes that require human determination" (Katsh & Rabinovitch-Einy, 2017, p. 70).

As technology evolves and disputes increase, the vacuum between disputes and their effective resolution can only grow. It seems inevitable that ADR will eventually fail in this situation. In anticipation of this, Alibaba has taken interesting steps. On any given day, the company has hundreds of millions of disputes that need to be resolved one way or another (ibid., p. 15). This volume of disputes is clearly not suited to either traditional court systems or ADR. Even hiring the thousands of workers it would take to assist in clearing such a backlog would still require some sort of infrastructure and training. Alibaba's solution? Make the users adjudicators. According to Alibaba's news agency, Alizila (yes, it's so big that it has its own news agency), a "user dispute resolution centre [was launched in 2014 with] over 800,000 people applying to become 'dispute assessors'."[1]

This is an example of community arbitration and dispute resolution employed on a massive scale, and consists of four levels—an initial automated level where the buyer can reach out to the seller to ask for a settlement; if this fails, the user can move to the user-based adjudication system described above; the third and final tier takes the client to a customer service representative. This matches something like the court system, where decisions can be appealed at higher and higher levels if a person is dissatisfied (and willing) to pursue the matter to another level. The innovation in Alibaba's case is that they provide a strong disincentive for customers to move to internal customer service in the form of lowered platform reputation levels for both buyer and seller (Katsh & Rabinovitch-Einy, 2017, p. 66).

---

[1]   https://www.alizila.com/alibaba-allows-users-to-play-judge-in-e-commerce-disputes/

Should the justice system think about entertaining and resolving disputes in a similar way? Various AI and computer-assisted interventions in the judicial system promise to make things better all-round for the pursuit of justice, but these same interventions can just as easily backfire, presenting to disputants biased, unfair, or just plain incorrect decisions on the basis of poorly stored and communicated information. Moreover, in a world with artificially intelligent entities in the form of bots scraping the internet for our data, all the way to chat agents that will replace human call center representatives, the legal field will become even more complex, as questions of culpability and retribution will no longer rest solely with human beings, but could extend to the realm of artificially intelligent entities as well. In summary, the future world of increased artificial intelligence, massive information surveillance and storage, and an exponential increase in network traffic and transaction volume will pose a heretofore unimaginable burden on the judicial systems of all countries and will require some form of technological innovation to mitigate.

Given the judicial institutions' historical inability to innovate in large part due to their very nature and function, the rise of AI in all its many facets (many of which have been discussed in previous chapters) should give us pause. If it took almost 500 years for the notion of patent and intellectual property law to evolve from its initial beginnings (MacLeod & Nuvolari, 2006), one should hold out little hope for a sudden change of pace now that technology is becoming even more sophisticated.

The field of legality and jurisprudence in a world populated with AI agents is now no longer so straightforward or easily decidable—not that many legal cases are, of course. Our point here is that things become hypercomplicated in a world that promises to be populated by elder care robots and self-driving cars. It is one thing to lend your car to someone else and have them get into an accident with it. It is a wholly other thing to lend your self-driving car to someone else and have the car get into the accident—in this latter situation, who is at fault? The car, or the "driver/passenger"? And who is the appropriate target for dispute resolution? The owner of the car? Or the programmer of the AI? Or the AI itself? And if the AI itself is to be blamed for something (in the same way, say, corporations are found guilty of some form of negligence today), how would justice be administered against them? Can one conceive of an AI jail, a virtual "sandbox" where an AI would spend some trillions of CPU cycles to "serve their time?" Or would the company agree to confine or erase their application and files?

Chapter 10

Incomplete Transfer (Upload or Die)

Previous chapters have looked at plausible ways in which AI could develop in the future. Throughout, we have argued that it isn't so much the question of whether AIs will be *actually* sentient and conscious in the future, but rather how they may be "taught" or "trained" using models created by humans. This implicit training will become, over time, a real feature of our world in the future, just as real as the roads, hydro wires, and other elements of infrastructure we depend on daily. But why is it that we are so confident machines will never be sentient or conscious? What is so special about mind, sentience, "the felt awareness of the subjective," that makes us so sure that AI in whatever form will never be able to encapsulate these tricky concepts?

To answer these questions, let us tackle the problem from the other vantage point. Rather than consider whether a machine can become conscious, let's entertain the quite popular idea of whether our consciousness could somehow be "transmuted" and put into a machine. Superficially, without any familiarity with the subject, one would not be blamed for thinking such a consideration to be absurd at the outset, and therefore not worth looking into. But a number of very brilliant individuals have argued the opposite using a seemingly quite compelling chain of reasoning.

The central assumption in this chain of reasoning is the theory of functionalism, described in the earlier chapter "Modeling the Mind." As we recall, functionalist theories of mind emphasize the results or outcomes in terms of behavior given a set of stimuli, whereas neurobiological theories of mind focus on the unique details of the anatomy and physiology of the brain and body that allow these working in combination to permit consciousness and intelligent behavior.

Chalmers (2010) takes quite seriously the proposal that consciousness could somehow be uploaded to some sort of computational medium. In pursuit of this conclusion, he too distinguishes between biological and functional theories of consciousness, writing that:

> "Biological theorists of consciousness hold that consciousness is essentially biological and that no nonbiological system can be conscious... [whereas] functionalist theorists of consciousness hold that what matters to consciousness is not biological makeup but causal structure and causal role, so that a nonbiological system can be conscious as long as it is organized correctly."

Chalmers admits that we are currently at a standstill in terms of research to understand how it is that a biological system can give rise to consciousness. However, he also points out that there does not appear to be any fundamental difference between what biological systems and nonbiological systems can give rise to in terms of causal effects. To argue the opposite is to move to some form of "vitalism," a long-discredited idea that there is some special "life essence" that animates all living things and disperses or is conserved but moves on upon the death of some living being. From a materialist perspective, biologists know that the bodies of living things are all made of the same sorts of atoms that non-living things are — there is no difference in principle at the atomic level between what living bodies are made of and what rocks, stars, or plastics are made of. Indeed, as some scientists caught in mystical reveries have reminded us, "we are made of star-stuff." Given such an unobjectionable and settled fact, the onus is on the biological theorist of consciousness to provide an account of just why biological systems are capable of consciousness whereas non-biological systems are not.

Functionalism is also compelling for the fact that it takes seriously the metaphor of "brain as computer." Note how this is described as a "metaphor," even though many cognitive scientists and philosophers have completely overlooked the metaphorical roots of this equivalence and treat it more or less as an article of faith that this is an accurate way of characterizing the role and function of the brain. This metaphor can be traced back to 1976 with the publication of Newell & Simon's "physical symbol system hypothesis." In this seminal paper, the authors claim that "a physical symbol system [like a digital computer] has the necessary and sufficient means for intelligent action."

Contrary to Chalmers' belief in the sufficiency of functionalism in capturing consciousness is Christof Koch. In his book, *The Feeling of Life Itself,* Koch does not mince words. His definition of consciousness? "Consciousness is experience" (2019, p. 1). Full stop. Taken another way, we are conscious when we experience things. And if we are to be able to "upload" our consciousness such that we can continue to live on in the virtual worlds of servers in the future, it will necessitate that we somehow also "experience" those virtual worlds in a way analogous to how we experience this world. Conversely, for machines to be conscious, they would need to experience our world in a way analogous to how we ourselves do.

As we explored in Chapter 3, according to Koch, there are five properties of consciousness. It is: private (I can know I am conscious but can't prove you are); structured (it contains separable objects); informative (it tells me something about myself and the world all the time); integrated (it presents a single picture of the world); and definite (it is inconceivable that more than one instance of experience could occur at one time).

Assuming for a moment that the dichotomy between functionalism and biological theories of consciousness is exhaustive of the possibilities, and assuming that a simulation of a biological brain would lead to the same result (i.e. the "emergence" of conscious experience of some unspecified kind), then it would indeed seem, in accordance with Chalmers, that functionalism is the only game in town when it comes to understanding how our own consciousnesses would be able to move from the biologically embedded "wet world" of carbon, oxygen, and hydrogen into the "dry world" of silicon and electricity. These, however, are two very major assumptions that need further investigation. These assumptions can be unpacked in terms of a set of questions.

Is the Activity of the Brain a <u>Necessary</u> Condition for Consciousness?

For the sake of argument, we shall assume indeed that the brain (in proper working order and being provided blood flow by the heart and oxygenation by the lungs) is a necessary condition for consciousness. It is instructive, however, to consider counterarguments to this assumption using veridical reports of near-death experiences in which the experiencer was able to accurately report third-person verifiable information even in the absence of all brain activity. The literature on NDEs has recently been growing. Admittedly, much of the literature is anecdotal and has no means by which it can be independently confirmed. However, a subset of NDEs has been documented that appears to provide evidence that people who are clinically dead (i.e. their brains do not function for an interval of time, either intentionally or through trauma) nevertheless report veridical perceptions and possess memories during intervals in which their brains are, by all current measurements, inactive. One of the most dramatic examples of such accounts is the case of Pam Reynolds, who in 1991 had the misfortune of being diagnosed with a massive aneurysm near her brainstem. The surgical intervention proposed necessitated that her body functions be brought to a halt. The aptly named "Operation Standstill" saw Reynolds' body cooled to such a point that blood would not flow throughout her body, the reasoning being that the brain would stand a better chance of recovery post-operation with a longer window of no oxygen at such a low temperature. Pam was anesthetized, her eyes taped shut, and an EEG was attached to monitor her cerebral cortex, the 'seat' of consciousness according to almost all neurological theories. Moreover, tiny speakers were placed in Reynolds' ears that emitted repeated 100 decibel clicks (the sound volume of a nearby train) to eliminate outside sounds.

While the surgical team was engaged in this laborious procedure, Pam subjectively felt herself "pop" out of the top of her head and was able to survey the operating theatre. She accurately identified the bone saw used to cut open her skull, conversations had between members of the team referring to the size of her arteries on one side of her body, and the odd nature of the haircut she had been given in preparation for the surgery. The entire procedure, from anesthesia to recovery, lasted some 7 hours (Beauregard, 2012).

Other independently corroborated veridical perceptions occurring during a time when the patient is clinically dead have also been recorded in the literature (Ring & Lawrence, 1993; Smit, 2008). A discussion of these cases, while fascinating, would take us too far afield in our discussion of the possibility of instantiating consciousness in computers or machines. And, as already mentioned, we will grant for the sake of argument that the proper functioning of a normal human brain is a necessary precondition for consciousness. However, these cases do suggest that there seem to be limit cases that occur in which the functioning of a brain is NOT necessary for consciousness (i.e. to use Koch's definition… "experience" to occur in a human subject, whatever the boundaries and parameters of such a subject might yet turn out to be).

Is the Activity of the Brain a *Sufficient* Condition for Consciousness?

In other words, given a perfectly functioning brain, is this *enough* to ensure that some conscious experience can be had by the subject? Once again, for the sake of argument, we will grant this claim as well, although here too there are limit cases. We should keep these in mind to ensure the right amount of humility in discussing the connection between brain and mind, as well as the presumed success that a recreation of brain states in some sort of simulation would inevitably lead to some form of consciousness/experience. To cast doubt on this assumption is a much simpler matter — in the case of routine anesthesia, for example, it seems clear that most people fall into some sort of state of unconsciousness that allows surgeons to perform operations of the most traumatic kind. If it weren't for the fact that the patient was blissfully unaware of such an operation taking place, they would have to endure the unimaginable torment of being operated on while awake. Understanding how anesthesia works to more or less "erase" consciousness during the interval of an operation is still (unsurprisingly, given how much we know about consciousness) a "work in progress." We do know that the first anesthetic (sulphuric ether) was used in 1846, and since then a whole pharmacopeia of chemicals has been discovered that perform the salutary effect of rendering a person unconscious while the nasty business of healing gets done—think amputations of limbs, opening of chest cavities, and introductions of foreign materials like pacemakers and shunts into the body. Therefore, while anesthesia seems to indicate situations where consciousness can be interrupted even though the brain continues to operate, we will assume that under most normal circumstances the brain is sufficient for consciousness.

Are All Brain Processes Computational in Nature?

If all brain processes were purely computational, and were accurately definable as such, we could conceive of a model that would be able to emulate (or simulate) a brain. However, details matter here. Using current weather simulations, meteorologists are able to simulate various trajectories of hurricanes and tropical storms to determine where they might make landfall or cause damage. They even give them names like Katrina or Bob. However, they are only able to approximately track the storm, and only at a certain level of detail.

Tiny, seemingly inconsequential changes to initial inputs of storm models can lead them to quite different predictions.

Now consider an actual human named Katrina or Bob (or Neo) who desires to be uploaded and simulated in some kind of Matrix. Consider also that, at the most conservative estimates, our brains have around 85 billion neurons, 85 trillion synapses (the junctions between neurons), and around 765 billion glial cells (Kassan, 2016). Glial cells used to be thought to be a purely structural component, something like mortar holding bricks together. However, we now seem to have found that the glial cells themselves have a kind of input on the function of the neurons. We still haven't mentioned axon and dendrite configuration. Now consider that, in order to simulate a hurricane so as to predict where it is going to go, there are at least around thirty predictive models with a number of subtle differences whose main job is to predict 1) how much rain a hurricane holds; 2) what direction it is going in; 3) what speed it is going at; 4) what its wind speed is; and 5) where it will cause damage. This is done using large and relatively easy to measure things like air pressure, temperature, water temperature, and even things like sun activity and seasonal temperature variation.

We know we are able to predict something like, or similar to, for the purposes of sending evacuation orders, where Katrina or Bob might end up and how much damage it might do. But even with a number of simplifying assumptions, the models we have to predict hurricanes are unable to exactly capture the "personality" of the hurricane. Why does this matter? Well, for hurricanes, it doesn't matter that much as long as you can get the general details right. But imagine signing up for a brain freezing or plasticizing procedure prior to having your precious organ cross-sectioned and uploaded to a server and being told that the model they would be using to capture your neural architecture may or may not be able to exactly capture "the essence of you." How close would be close enough? 75%, 80%, 60%? What if the upload team were unable to give you accurate percentages because of the very small chaotic differences that the initial states of the models would have to their outcomes? So even if we grant for the sake of argument that brain operation is exhausted using some computational theory of mind, we nevertheless have to deal with the sheer number and complexity of parameters that have to be put into the model to get anything like YOU to come out the other end.

Is a Simulation the Same as the Real Thing?

In a previous chapter ("The Importance of Embodied Human Intelligence"), we mentioned the concept of the video game *The Sims*. These are simulated people that move around in a simulated neighborhood pursuing simulated goals. They drink simulated water and eat simulated food to quell simulated appetites. There is a reason we refer to all of these things as being simulated—because there is a strict ontological difference between what is simulated and what is real. If there were no such difference, these words would be synonyms. They're not. A simulated fire in the world of *The Sims* cannot make your actual home burn down, no matter how long you leave your computer on. Running a simulation of a hurricane in your neighborhood will not cause it to flood and will not necessitate your evacuation. A simulated airplane crash will carry with it no fatalities and will not be reported on the evening news.

Why do so many cognitive scientists and neuroscientists somehow think that the brain is different from these examples of simulated unrealities? We think it has something to do with the nebulous and intangible notions of consciousness and intelligence, as well as a confusion between the kinds of nebulosity and intangibility that these two quite different phenomena possess. For instance, while a simulated hurricane will not get you wet, a simulated chess board with your poor king helplessly backed in the corner by the likes of a StockFish or AlphaZero WILL make you lose at that game of chess. The instantiation of the chess board and its pieces matters little — you could play the game by getting instructions typed move by move to you from inside the bowels of the chess algorithm and play the game on a physical board. Because of the fact that intelligence and intelligent reasoning seem to go so closely hand in hand with the existence of consciousness, it seems natural and reasonable to think the same of consciousness. However, we maintain that this is a deceptive similarity and that there is no sense in which StockFish or AlphaGo "enjoys" its victories in the same way that a human might. Therefore, the converse would also hold: even if the structure of your brain were simulated to an arbitrary degree of accuracy, it would make no difference to you, for "you" would no longer be there.

The Problem of Other Minds
— How Can We Ever Know that Others
(Carbon or Silicon Based) Are Conscious?

Let's say we get all of the details right. We've used an incredibly precise model of a brain. We've taken detailed samples of the behavior of YOUR brain in particular, so we can match neural outputs to inputs. We've sat down with you and reviewed the process, and gotten you to sign a waiver saying that you do not hold us responsible in case our model does not lead completely to a full recreation of your subjectivity. You enter the operating room and undergo the "process," whatever it may turn out to be.

The doctors and scientists complete the process. Your body (sans brain) is disposed of, no longer necessary after this groundbreaking procedure. The medical team huddles around an interface with clear indications that the operation was a success.

How do we really know that it was a success? The team has already thought of this little conundrum and has agreed with you beforehand that they will ask you a set of "challenge" questions after you have been uploaded, much like the questions you might get asked if you forget your password, like "What is your favorite color?" or "Who was your favorite teacher?"

The team might get some replies come up on their computer screen, or even have some emulation of your voice to give them responses. But would the team really know that a "person" like you was now inside this digital utopia? How would they have known there was a "you" before they did the procedure? This is an ineliminable epistemological challenge that absolutely no one is able to address, regardless of how confident they may be with the technological wizardry that would enable such an operation and the faith they have in the computational theory of mind.

Summary

We have discussed above the daunting task of trying to capture all of the various complexities and functional parts of the brain. Despite the seeming intractability and complexity of the task, there are still a number of scientists and engineers who think all of our objections outlined above are merely negative thinking —a case of "luddites" who would rather put their heads in the sand than face the future realities of digital immortality. In the preface of this book, we mentioned three camps of AI opinion, characterizing Ray Kurzweil as a "digital utopian" using Max Tegmark's classification. A quote from his 2005 book, *The Singularity is Near: When Humans Transcend Biology*, is worth considering, just to see how idealistic he is about solving these fundamental problems:

> "...the end of the 2030s is a conservative projection for successful uploading. We should point out that a person's personality and skills do not reside only in the brain... Our nervous system extends throughout the body, and the endocrine (hormonal) system has an influence, as well... Confirmation of the uploading milestone will be in the form of a "Ray Kurzweil" or "Jane Smith" Turing test, in other words convincing a human judge that the uploaded re-creation is indistinguishable from the original specific person... However, uploading the nonbiological portion of intelligence will be relatively straightforward, since the ease of copying computer intelligence has always represented one of the strengths of computers." (Kurzweil, 2005, p. 177)

As described in the thought experiment above, no amount of "Turing testing" will ever convince us that a simulation of a person is conscious—just like no method currently exists (other than heuristic inference through embodied empathy of creatures who are relatively like us) to determine if our flesh and blood neighbors are conscious. Having machines succeed at password challenges should not be enough to convince anyone (let alone the person undergoing the brain scanning in order to be uploaded) that their subjectivity will be intact on the other side of the firewall.

Another scientist who sees all the difficulties we have outlined above as mere "engineering problems" is Ken Hayworth, writing in his article "Mind Uploading" in 2016:

> "If the brain's functioning is governed by the causal laws of physics then any subset of the brain's neurons should in principle be replaceable by a computer simulation of those same neurons hooked up to the rest with electrodes. As long as the causal relationships are maintained then the outward behavior of the person must remain the same, even to the extent of verbally claiming to have the same conscious experiences." (p. 18)

He calls this "the slippery slope of materialism." Slippery slope, indeed. What warrants Hayworth's confidence that a preservation of causal relationships between a simulation and reality would thereby preserve the subject intact? Like the simulated fire that does not burn, and the simulated food that does not sate, simulated consciousness does not experience. What would lead one to think anything otherwise? Capturing the causal connections between speed of wind, pressure, direction, and solar energy in a model does not conjure a hurricane—why would it be different for consciousness?

Kassan (2016) makes some points that demonstrate the entire "uploading of your mind" question to be, at best, incoherent. Central to his argument is the idea for "uploaders" to have some digital and discrete means by which our mind could be represented. After all, computers are "digital" machines, meaning that they only work for fixed values. The problem right off the bat, and a bit of introspection into your own mind will reveal it, is that our mind uses words and intuitions to deliberate, reflect, daydream, or come to conclusions about the world and the situation we find ourselves in.

Moreover, these words are anything but fixed and rigid designators of the various conceptual objects we entertain in our mind's eye. Kassan points out that, whereas the mind is an intangible beast due to this feature, and therefore probably not amenable to a straightforward "upload," we DO know that brains are always found in the company of minds, right? And so that should mean that if we copy the brain in all its detail and then stick it on a server somewhere, surely that will do the trick, won't it? Well, who knows…

In keeping with previous chapters, we have argued that a) sentience and consciousness depends on some sort of embodiment; b) the embodiment of consciousness is not limited only to the body, but also to the environment in which it is situated; c) the self is best understood as a limiting case of the entire self/ environment matrix in which the self finds itself embodied; d) therefore, the self is i) unique and yet paradoxically ii) shared by all instances of self that exist in the world.

In agreement with Roger Penrose (*Shadows of the Mind, The Emperor's New Mind*), there is something **non-computational** about consciousness that makes it utterly resistant to current approaches being used in AI to simulate intelligent behavior. What this also means, as a corollary, is that there will never be a way in which we would be able to upload our consciousness to some sort of computational matrix, as we are unable to specify what exactly the computational procedure would be to do so. In other words, it is not a mere matter of simulating enough connections between neurons, or even a matter of "embodying" the resulting simulation in some sort of machine that interacts with an environment, for the simple reason that we just would not know where or how to begin with such an exercise, nor how to confirm that we had done it correctly.

We should not hold out hope that our minds can be uploaded to computers. Our identities are not computational and mere simulations of our identities (whatever this would even mean) would not be able to "conjure" consciousness or identity into existence. Alas, our hopes for immortality will have to lie elsewhere. In the meantime, our ongoing journey with AI and digital technology will show us more about becoming artificial.

References

A Brief History of
Artificial Intelligence

Adams, R. L. (2017) 10 Powerful Examples of Artificial Intelligence in Use Today, *forbes.com*, 10 January, [Online], https://www.forbes. com/sites/robertadams/2017/01/10/10-powerful-examples-of-artificial-intelligence-in-use-today/#1627b200420d [29 August 2019].

Bacon, Sir F. (1620/2000) The New Organon: Novem Organum Scientiarum, in *Francis Bacon: The New Organon*, Cambridge Texts in the History of Philosophy, Cambridge: Cambridge University Press.

Colman, A. (2006) *A Dictionary of Psychology*, p. 670, New York: Oxford University Press.

Dua, A. (2019) Artificial Intelligence Applications: Retail Use Cases, *dataversity.net*, 4 April, [Online], https://www.dataversity.net/ artificial-intelligence-applications-retail-use-cases/ [24 August 2019].

Foote, K.D. (2016) A Brief History of Artificial Intelligence, *dataversity.net*, 5 April, [Online], https://www.dataversity.net/ brief-history-artificial-intelligence/ [24 August 2019].

Giles, T. (2016) Aristotle Writing Science: An Application of His Theory, *Journal of Technical Writing and Communication*, 2016, pp. 83–104.

Lewis, T. (2014) Brief History of Artificial Intelligence, *livescience.com*, 4 December, [Online], https://www.livescience.com/49007-history-of-artificial-intelligence.html [26 August 2019].

Lighthill Report (1973) [Online], http://www.chilton-computing. org.uk/inf/literature/reports/lighthill_report/contents.htm   [9 February 2020].

Long, L.N. (2016) Toward Human-Level Massively-Parallel Neural Networks With Hodgkin-Huxley Neurons, in Steunebrink, B., Wang, P. & Goertzel, B. (eds.) *Artificial General Intelligence. AGI 2016: Lecture Notes in Computer Science*, vol. 9782, Cham: Springer.

McCorduck, P. (1979) *Machines Who Think*, 2nd ed., Natick, MA: A.K. Peters, Ltd.

McKeon, R. (ed.) (1941) *The Organon*, Oxford: Random House with Oxford University Press.

O'Connor, K.M. (1994) *The Alchemical Creation of Life (Takwin) and Other Concepts of Genesis in Medieval Islam*, PhD dissertation, University of Pennsylvania.

Popple, J. (1993) *SHYSTER: A Pragmatic Legal Expert System*, PDF, PhD thesis, Australian National University, [Online], http://dx.doi. org/10.2139/ssrn.1335637 [30 May 2009].

Russell, S.J. & Norvig, P. (2003)*Artificial Intelligence: A Modern Approach*, 2nd ed., Upper Saddle River, NJ: Prentice Hall.

Schaeffer, J. (1997/2009) *One Jump Ahead: Challenging Human Supremacy in Checkers*, New York: Springer.

Smith, C., McGuire, B., Huang, T. & Yang, G. (2006) The History of Artificial Intelligence, *courses.cs.washington.edu*, December, [Online], https://courses.cs.washington.edu/courses/csep590/06au/projec ts/history-ai.pdf [27 August 2019].

Turing, A. (1950) Computing Machinery and Intelligence, *abelard.org*, [Online],   https://www.abelard.org/turpap/turpap.php   [28 August 2019].

Winston, P.H. & Prendergast, K.A. (1986) CADUCEUS: An Experimental Expert System for Medical Diagnosis, in *The AI Business: Commercial Uses of Artificial Intelligence*, pp. 67–80, Cambridge, MA: MIT Press.

Zang, Y., Zhang, F., Di, C. & Zhu, D. (2015) Advances of Flexible Pressure Sensors Toward Artificial Intelligence and Healthcare Applications, *pubs.rsc.org*, [Online], https://pubs.rsc.org/en/content/articlehtml/2015/mh/c4mh00147h [28 August 2019].

The Importance of Embodied Human Intelligence

Blakeslee, S. & Blakeslee, M. (2008) *The Body Has a Mind of Its Own: New Discoveries about How the Mind-Body Connection Helps Us Master the World*, New York: Random House.

Kassan, P. (2016) Uploading Your Mind Does Not Compute, *Skeptic*, 21 (2), pp. 24–27.

Kriss, S. (2017) You Think With the World, Not Just Your Brain, *The Atlantic*, 13 October, [Online], https://www.theatlantic.com/science/archive/2017/10/extended-embodied-cognition/542808/ [23 March 2020].

Kurzweil, R. (2014) *How to Create a Mind: The Secret of Human Thought Revealed*, London: Duckworth.

McNerney, S. (2011) A Brief Guide to Embodied Cognition: Why You Are Not Your Brain, *Scientific American Blog*, 4 November, [Online], https://blogs.scientificamerican.com/guest-blog/a-brief-guide-to-embodied-cognition-why-you-are-not-your-brain/ [10 February 2020].

Melnick, A. (2011) *Phenomenology and the Physical Reality of Consciousness*, Amsterdam: John Benjamins.

Merleau-Ponty, M. (1962) *Phenomenology of Perception*, London: Routledge.

Naskar, A. (2018) Pretending to Have Sentience is not the Same as Having Sentience: On Artificial Intelligence, *Naskarism*, 6 December, [Online], https://naskarism.wordpress.com/2018/12/06/pretending-to-have-sentience-is-not-the-same-as-having-sentience-on-artificial-intelligence/ [12 July 2019].

Modeling the Mind

Aaronson, S. (2014) *Why I Am Not An Integrated Information Theorist (or, The Unconscious Expander)*, [Online] https://www.scottaaronson.com/blog/?p=1799 [12 June 2020].

Cerullo, M.A. (2015) The Problem with Phi: A Critique of Integrated Information Theory, *PLoS Computational Biology*, 11 (9), e1004286, [Online], https://doi.org/10.1371/journal.pcbi.1004286.

Churchland, P.S. (1986) *Neurophilosophy: Toward a Unified Science of the Mind-Brain: Computational Models of Cognition and Perception*, Cambridge, MA: MIT Press.

Clark, A. (1989) *Microcognition: Philosophy, Cognitive Science, and Parallel Distributed Processing*, Explorations in Cognitive Science, 6, Cambridge, MA: MIT Press.

Clark, A. (2001) *Mindware: An Introduction to the Philosophy of Cognitive Science*, New York: Oxford University Press.

Fodor, J.A. & Pylyshyn, Z.W. (1988) Connectionism and Cognitive Architecture: A Critical Analysis, *Cognition*, 28 (1–2), pp. 1–2.

Fromkin, V.A. (2000) *Answer Key for Linguistics: An Introduction to Linguistic Theory*, Malden, MA: Blackwell.

Kim, J. (2005) *Mind as a Computer: Machine Functionalism*, pp. 76–77; 80–96, Boulder, CO: Westview Press.

Koch, C. (2019) *The Feeling of Life Itself. Why Consciousness is Widespread but Can't Be Computed*, Cambridge, MA: MIT Press.

Marsh, H. (2019) Can Man Ever Build a Mind?, *ft.com*, 10 January, [Online], https://www.ft.com/content/2e75c04a-0f43-11e9-acdc-4d9976f1533b [12 July 2019].

Naskar, A. (2018) Pretending to Have Sentience is not the Same as Having Sentience: On Artificial Intelligence, *Naskarism*, 6 December, [Online], https://naskarism.wordpress.com/2018/12/06/pretending-to-have-sentience-is-not-the-same-as-having-sentience-on-artificial-intelligence/ [12 July 2019].

Searle, J.R. (1992) *The Rediscovery of the Mind: Representation and Mind*, Cambridge, MA: MIT Press.

Tononi, G. (2004) An Information Integration Theory of Consciousness, *BMC Neuroscience*, 5, 42, [Online], https://doi.org/10.1186/1471-2202-5-42 [11 June 2020].

The Changing Nature of War

Bergen, P. & Tiedemann, K. (2010) *The Year of the Drone: An Analysis of U.S. Drone Strikes in Pakistan, 2004–2010*, Washington, DC: New American Foundation.

Dixon, J.D.R. (2000) *UAV Employment in Kosovo: Lessons for the Operational Commander*, Newport, RI: Naval War College..

Grier, P. (2009) Drone Aircraft in a Stepped-up War in Afghanistan and Pakistan, *The Christian Science Monitor*, [Online], http://www.csmonitor.com/USA/Military/2009/1211/Drone-aircraft-in-a-stepped-up-war-in-Afghanistan-and-Pakistan [8 July 2011].

Wagner, W. (1982) *Lightning Bugs and other Reconnaissance Drones: The Can-do Story of Ryan's Unmanned Spy Planes*, Fallbrook, CA: Aero Publishers.

Zaloga, S.J. & Palmer, I. (2008) *Unmanned Aerial Vehicles: Robotic Air Warfare 1917–2007*, New York: Osprey.

Acting

Ebiri, B. (2019) Even Now, Rutger Hauer's Performance in 'Blade Runner' is a Marvel, *New York Times*, 25 July, [Online], https://www.nytimes.com/2019/07/25/movies/blade-runner-rutger-hauer.html [15 September 2019].

Genzlinger, N. (2018) Douglas Rain, 90, Shakespearean and Voice of Computer Named HAL, Dies, *New York Times*, 12 November, [Online], https://www.nytimes.com/2018/11/12/obituaries/douglas-rain-dead.html [17 September 2019].

O'Carroll, E. & Driscoll, M. (2018) '2001: A Space Odyssey' Turns 50: Why HAL Endures, *Christian Science Monitor*, 3 April, [Online], https://www.csmonitor.com/Technology/2018/0403/2001-A-Space-Odyssey-turns-50-Why-HAL-endures [15 September 2019].

Wickman, F. (2012) I, Actor: Cinema's Finest Robot Performances, *slate.com*, 9 June, [Online], https://slate.com/culture/2012/06/michael-fassbenders-david-in-prometheus-is-just-the-latest-great-robot-performance.html [10 September 2019].

Wood, M. (2017) Michael Fassbender Reveals the Differences Between His Covenant and Prometheus Characters, *cinemablend.com*, 23 April, [Online], https://www.cinemablend.com/news/1650389/michael-fassbender-reveals-the-differences-between-his-covenant-and-prometheus-characters [14 September 2019].

Twentieth Century Fox Home Entertainment (2017) *Alien: Covenant.*

Seeking Succor in Sentience

Abdi, J., Al-Hindawi, A., Ng, T. & Vizcayshipi, M.P. (2018) Scoping Review on the Use of Socially Assistive Robot Technology in Elderly Care, *BMJ Open*, [Online], https://bmjopen.bmj.com/content/8/2/e018815 [6 September 2019].

Anderson, M. & Perrin, A. (2017) Tech Adoption Climbs Among Older Adults, *Pew Research Center*, 17 May, [Online], https://www.pewinternet.org/2017/05/17/tech-adoption-climbs-among-older-adults/?xid=ps_smithsonian [1 September 2019].

Holland, P. (2018) Fur-Covered Robots are the New Therapy Animal, *cnet.com*, 4 April, [Online], https://www.cnet.com/news/comfort-robots-aflac-duck-and-paro-seal-help-sick-people/ [6 September 2019].

Kahn, P.H. Jr. & MacDorman, K.F. (2007) Psychological Benchmarks of Human-Robot Interaction, *Interaction Studies: Social Behaviour and Communication in Biological and Artificial Systems*, 8 (3).

Satariano, A., Peltier, E. & Kostyukov, D. (2018) Meet Zora, the Robot Caregiver, *New York Times*, 23 November, [Online], https://www.nytimes.com/interactive/2018/11/23/technology/robot-nurse-zora.html.

Statt, N. (2017) The Leka Smart Toy is a Robot for Children with Developmental Disabilities, *The Verge*, 4 January, [Online], https://www.theverge.com/ces/2017/1/4/14167590/leka-smart-toy-robot-autism-learning-tool-ces-2017 [6 September 2019].

Tarantola, A. (2017) Robot Caregivers are Saving the Elderly from Lives of Loneliness, *engadget.com*, 29 August, [Online], https://www.engadget.com/2017/08/29/robot-caregivers-are-saving-the-elderly-from-lives-of-loneliness/ [3 September 2019].

Wachsmuth, I. (2018) Robots Like Me: Challenges and Ethical Issues in Aged Care, *National Institutes of Health*, 3 April, [Online], https://www.ncbi.nlm.nih.gov/pmc/articles/PMC5892289/ [1 September 2019].

Alone at the Water Cooler

Altucher, J. (2020, January 2) Being worried that automation and AI will hurt society is like being worried in 1440 that the printing press would hurt society and cause mass unemployment of scribes, *Twitter*, [Online], https://twitter.com/jaltucher/status/1212586810603450369.

Computer History Museum (2020) Joseph Clement, *Computer History Museum*, [Online], https://www.computerhistory.org/babbage/josephclement/ [19 June 2020].

Encyclopedia Britannica (2019) Player Piano, *Encyclopedia Britannica*, [Online], https://www.britannica.com/art/player-piano [19 June 2020].

Hern, A. (2019) New AI Fake Text Generator May Be Too Dangerous to Release, Say Creators, *The Guardian*, [Online], https://www.theguardian.com/technology/2019/feb/14/elon-musk-backed-ai-writes-convincing-news-fiction [19 June 2020].

Hsu, F.-H. (2002) *Behind Deep Blue: Building the Computer that Defeated the World Chess Champion*, Chapter 12, pp. 204–258, Princeton, NJ: Princeton University Press.

King, A. (2020) *Talk to Transformer* [computer application using GPT-2 language model], [Online], https://talktotransformer.com/ [16 June 2020].

Munro, A. (2012) *Dear Life*, Toronto: Douglas Gibson/McLelland & Stewart.

Sadler, M. & Regan, N. (2019) *Game Changer: AlphaZero's Groundbreaking Chess Strategies and the Promise of AI*, Chapter 5, p. 89, Alkmaar: New In Chess.

Standage, T. (2002) *The Mechanical Turk: The True Story of the Chess-Playing Machine that Fooled the World*, London: Allen Lane.

Zuboff, S. (2019) *Surveillance Capitalism: The Fight for a Human Future at the New Frontier of Power*, New York: Public Affairs.

Do As You Are Told

Borges, J. (1964) The Analytical Language of John Wilkins, *Other Inquisitions: 1937–1952*, Austin, TX: University of Texas Press.

Schopenhauer, A. (1969) *The World as Will and Representation*, vols. 1 & 2, New York: Dover Publications.

Young, S. (nd) T w e n t y S i d e d, *shamusyoung.com*, [Online], https://www.shamusyoung.com/twentysidedtale/?p=7556.

The Devil Made Me Do It

Bontke, J. (2019) Austin Couple Plans to Sue AirBnB After Guests Damaged Homes, *cbsaustin.com*, [Online], https://cbsaustin.com/news/local/austin-couple-plans-to-sue-airbnb-after-guests-damaged-homes.

Bourne, J. (2020) Airbnb Uses AI-Enabled Trait Analyser to Check if its Customers are Psychopaths, *AI News*, [Online], https://artificialintelligence-news.com/2020/01/09/airbnb-uses-ai-enabled-trait-analyser-to-check-if-its-customers-are-psychopaths/ [12 June 2020].

Brenner, S. (2007) *Law in an Era of 'Smart' Technology*, New York: Oxford University Press.

Collinson, P. (2018) What Happens When AirBnB Goes Wrong, *The Guardian*, 4 August, [Online], https://www.theguardian.com/technology/2018/aug/04/what-happens-when-airbnb-goes-wrong.

DeBronkart, D. (2009) Imagine Someone Had Been Managing Your Data, and Then You Looked, *participatorymedicine.org*, [Online], https://participatorymedicine.org/epatients/2009/04/imagine-if-someone-had-been-managing-your-data-and-then-you-looked.html [12 March 2020].

Dubminsky, Z., Ouellet, V. & Forero, L. (2019) Airbnb Quietly Shut Down a Top Host Amid Scathing Reviews, but Hundreds of Guests Were Left to Stay with Him, *CBC*, [Online], https://www.cbc.ca/news/canada/montreal/airbnb-montreal-aj-host-suspended-accounts-1.5252233

Katsh, M. & Rabinovitch-Einy, O. (2017) *Digital Justice: Technology and the Internet of Disputes*, Oxford: Oxford University Press.

Kafka, F. (1925) *The Trial*, London: Vintage.

MacLeod, C. & Nuvolari, A. (2006) Inventive Activities, Patents and Early Industrialization: A Synthesis of Research Issues, No 06–28, *DRUID Working Papers, DRUID*, Copenhagen Business School,

Department of Industrial Economics and Strategy/Aalborg University, Department of Business Studies, [Online], https://EconPapers.repec.org/RePEc:aal:abbswp:06-28 [20 June 2020].

Mills, J. (2019) AirBnB Guests Trash Mansion Causing £445,000 Damage, *Metro*, 31 December, [Online], https://metro.co.uk/2019/12/31/airbnb-guests-trash-mansion-causing-445000-damage-11980104/

Padojino, J. (2019) Party at AirBnB Rental Goes Awry When Three Men Rob Property Owner, *paloaltoonline.com*, 2 October, [Online], https://www.paloaltoonline.com/news/2019/10/02/party-at-airbnb-rental-goes-awry-when-three-men-rob-property-owner.

Temperton, J. (2020) Threats, Fear and Chaos: The Messy Fall of an AirBnB Scam Empire, *Wired*, [Online], https://www.wired.co.uk/article/airbnb-scam-london-suspended.

Incomplete Transfer (Upload or Die)

Beauregard, M. (2012) Near-Death Explained, *salon.com*, 21 April, [Online], https://www.salon.com/2012/04/21/near_death_explained/ [19 March 2020].

Chalmers, D.J. (2010) Mind Uploading: A Philosophical Analysis, in Blackford, R. & Broderick, D. (eds.) *Intelligence Unbound: The Future of Uploaded and Machine Minds*, Oxford: Blackwell.

Hayworth, K. (2016) Mind Uploading: An Argument for the Scientific and Technical Plausibility of Preserving Thoughts Indefinitely, *Skeptic*, 21 (2), p. 18.

Kassan, P. (2016) Uploading Your Mind Does Not Compute, *Skeptic*, 21 (2), pp. 24–27.

Koch, C. (2019) *The Feeling of Life Itself: Why Consciousness is Widespread but Can't Be Computed*, Cambridge, MA: MIT Press.

Kurzweil, R. (2005) *The Singularity is Near: When Humans Transcend Biology*. New York: Penguin Books.

Newell, A. & Simon, H.A. (1976) Computer Science as Empirical Inquiry: Symbols and Search, *Communications of the ACM*, 19 (3), pp. 113–126.

Penrose, R. (1996) *Shadows of the Mind: A Search for the Missing Science of Consciousness*, London: Oxford University Press.

Penrose, R. & Gardner, M. (2016) *The Emperor's New Mind: Concerning Computers, Minds, and the Laws of Physics*, Oxford: Oxford University Press.

Ring, K. & Lawrence, M. (1993) Further Evidence for Veridical Perception During Near-Death Experiences, *Journal of Near-Death Studies*, 11, pp. 223–229.

Smit, R.H. (2008) Corroboration of the Dentures Anecdote Involving Veridical Perception in a Near-Death Experience, *Journal of Near-Death Studies*, [Online], https://digital.library.unt.edu/ark:/67531/metadc798921/ [28 March 2020].